Hartmut Bossel
System Zoo 2 Simulation Models
Climate, Ecosystems, Resources

AF190968

About the author: Hartmut Bossel is Professor Emeritus of environmental systems analysis. He taught for many years at the University of California in Santa Barbara and the University of Kassel, Germany, where he was director of the Center for Environmental Systems Research until his retirement. He holds an engineering degree from the Technical University of Darmstadt, and a Ph.D. degree from the University of California at Berkeley. With a background in engineering, systems science, and mathematical modeling, he has led many research projects and future studies in different countries, developing computer simulation models and decision support systems in the areas of energy supply policy, global dynamics, orientation of behavior, agricultural policy, and forest dynamics and management. He has written numerous books on modeling and simulation of dynamic systems, social change and future paths, and has published widely in the scientific literature in several fields.

SYSTEM ZOO 2 SIMULATION MODELS

Climate, Ecosystems, Resources

Hartmut Bossel

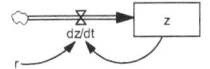

System Zoo 2 Simulation Models
© Hartmut Bossel 2007

Printed and published by
Books on Demand GmbH, Norderstedt, Germany

ISBN 978-3-8334-8423-0

Bibliografische Information Der Deutschen Bibliothek:
Die Deutsche Bibliothek verzeichnet diese Publikation
in der Deutschen Nationalbibliografie;
detaillierte bibliografische Daten sind im Internet über
http://dnb.ddb.de abrufbar.

Bibliographic information published by Die Deutsche Bibliothek:
Die Deutsche Bibliothek lists this publication
in the Deutsche Nationalbibliografie;
detailed bibliographic data are available in the Internet at
http://dnb.ddb.de

Information bibliographique de Die Deutsche Bibliothek:
Die Deutsche Bibliothek a répertorié cette publication
dans le Deutsche Nationalbibliografie;
les données bibliographiques détaillées peuvent être consultées
sur Internet à l'adresse http://dnb.ddb.de.

Preface

Daily life and the development of our world are determined by complex coupled dynamic systems: People, animals, plants, technology, enterprises, towns, rivers, forests. Although seemingly permanent in their outer appearance, they are constantly changing by mostly invisible processes, thereby also changing their environment. Knowledge about their possible dynamics is essential in many areas. The dynamic processes must be exposed and analyzed with the tools of systems analysis: with mathematical modeling and computer simulation.

The "System Zoo" is a collection of about one hundred simulation models[1] of complex dynamic systems from all areas of life in the departments: elementary systems, physics and engineering, climate and vegetation, ecosystems and resources, economy and society, global development. It is published in three volumes:

System Zoo 1 Simulation Models – Elementary Systems, Physics, Engineering
System Zoo 2 Simulation Models – Climate, Ecosystems, Resources
System Zoo 3 Simulation Models – Economy, Society, Development

All of these models are formulated according to globally accepted "system dynamics" standard, are documented in complete detail, have been thoroughly checked, are operational and can be operated with freely available simulation software with extensive processing possibilities. The model documentations are complemented by numerous exercises. The models are small enough to be implemented and utilized with modest effort, but most of them exhibit complex and often surprising behavior that is beyond intuitive assessment. Computer simulation provides a simple means of investigating and understanding the surprising variety of possible behaviors – a variety similar to that found in a zoo full of exotic animals. The "animals" in the System Zoo are grouped in the following six categories:

Part 1 "ELEMENTARY SYSTEMS" of the System Zoo documentation introduces smaller systems which are found as components in many complex systems, substantially determining their dynamics (like exponential and logistic growth, oscillations, delays etc.). This part is also an introduction to the practical side of modeling and simulation.

Part 2 "PHYSICS AND ENGINEERING" deals with an area in which the mathematical modeling of dynamic systems has arisen and in which simulations have always been quite important. The behavioral peculiarities of complex (nonlinear) systems such as limit cycles, attractors, multiple equilibrium points, and chaos are also examined here. More complex models are documented from the areas of control engineering, flight dynamics, and aerodynamics.

In Part 3 "CLIMATE AND VEGETATION" applications are introduced from the areas of climatology, the global CO_2 budget, photosynthesis and biomass production of

---

[1] Original publication: Bossel, H.: Systemzoo 1 – Elementarsysteme, Technik und Physik; Systemzoo 2 – Klima, Ökosysteme und Ressourcen; Systemzoo 3 – Wirtschaft, Gesellschaft und Entwicklung (3 vols.). Books on Demand, Norderstedt/Germany 2004. CD (German edition) containing all models: Bossel, H.: Systemzoo. coTec Verlag, Rosenheim/Germany 2005.

plants, forest growth, and water, energy, and nutrient budgets of plant production in agriculture.

Part 4 "ECOSYSTEMS AND RESOURCES" deals primarily with the dynamics resulting from interaction of plants, animals, and people with other organisms and environmental resources: by competition for food and nutrients, by use of renewable resources and exploitation of nonrenewable resources.

In Part 5 "ECONOMY AND SOCIETY" dynamic processes from this area are modeled and simulated: processes of production, supply chains, sales and consumption, competition for markets, personal life planning, unemployment, impacts of taxes on the development of commuter traffic and the economy, and finally also socio-psychological processes like escalation, dependence and aggression.

Part 6 "GLOBAL DEVELOPMENT" presents simulation models which are relevant for the examination of longer-term societal developments with respect to population, housing, cost of living, pensions, state indebtedness, globalization, international competition, and global long-term development. Nonnumerical knowledge processing for the simulation of complex decision processes and impact assessment is also introduced.

The System Zoo is a collection of results and experience from a large number of research projects and many years of teaching, model development, and computer simulation. It is particularly well suited for courses in modeling and simulation, for class work as well as self-study, and for projects in schools, colleges, and universities. The three volumes of the System Zoo are complemented by a companion book providing the theoretical and practical foundations of mathematical modeling and computer simulation of systems: Hartmut Bossel: *Systems and Models – Complexity, Dynamics, Evolution, Sustainability*. Books on Demand, Norderstedt 2007 (ISBN 978-3-8334-8121-5). More detail on dynamic systems is provided in an earlier text: Hartmut Bossel: *Modeling and Simulation*. A K Peters, Wellesley MA 1994 (ISBN 1-56881-033-4).

All System Zoo models (English versions only) are available for free download from the Center of Environmental Systems Research of the University of Kassel (http://www.usf.uni-kassel.de/cesr/).

Contents

Contents of other System Zoo volumes:

SYSTEM ZOO 1 SIMULATION MODELS

1 ELEMENTARY SYSTEMS

2 PHYSICS AND ENGINEERING

SYSTEM ZOO 3 SIMULATION MODELS

5 ECONOMY AND SOCIETY

6 GLOBAL DEVELOPMENT

Working with the System Zoo

Myriads of interconnected complex dynamic systems determine the development of our world and daily life. We recognize them in their particular and more or less permanent shape: people, animals, plants, forests, technology, enterprises, towns, states. But we hardly know and seldom recognize them as dynamic systems which are constantly changing by mostly invisible processes, thereby also changing their environment. This aspect usually escapes direct observation. It must be revealed using the tools of system analysis – in analogy to using the tools of x-ray analysis to reveal the structure and function of our bodies.

Animals and systems can be pictured and described in detail in encyclopedias and textbooks, but to get to know and understand their behavior we must observe them for some time under different conditions. Zoological gardens were created to provide people with the opportunity of observing animals – and in particular "strange" animals. In a zoo we can discover what books about animals cannot provide: the behavioral dynamics of a living being, often even in direct interaction with us. And a zoo offers a collection of very different animals with quite diverse behaviors in its different departments: mammals and birds, amphibians and fish, big and small animals, solitary and gregarious animals.

The three volumes of the System Zoo offer in their six chapters a collection of about one hundred simulation models of complex dynamic systems of all areas of life in the departments: elementary systems, physics and engineering, climate change and vegetation, ecosystems and resources, economy and society, global development. These simulation models are brought to life with easy-to-use simulation software. The models are described briefly at the beginning of each chapter. These descriptions should be read first to obtain an overview of the System Zoo and its residents.

The models and their simulation programs are all fully documented. They can be implemented on the PC with little effort using freely available interactive simulation software. It is the interactive work with these "system animals" that brings about often surprising insights about their dynamics and their often peculiar behavior. To save space, only one representative simulation run is usually documented for each model, but the spectrum of possible behaviors is always much richer than can be shown here. Each model description therefore mentions further interesting aspects which one should investigate to really understand the behavior of the system. Important: Most models are "generic" and are therefore valid also in completely different applications. Corresponding hints are provided in the respective model description.

As in a visit to a zoo one should initially concentrate on systems of greatest personal interest. If you are new to the field of simulation, start your excursion with some of the simpler systems from the chapter ELEMENTARY SYSTEMS to familiarize yourself with the simulation software and its many processing possibilities. Detailed documentation and many teaching examples are also provided with the simulation software. The models in the different chapters are largely independent of each other. It therefore is not necessary (and not recommended) to complete the models one after the other. Rather, let your own interest and the joy of investigating and observing strange "animals" guide you.

The simulation models were developed using the software Vensim PLE (Personal Learning Environment) which is freely available in the internet for teaching purposes and private use (http://www.vensim.com). The symbolism ("system dynamics") used here is also used by other common simulation software like Stella (or ithink, http://www.iseesystems.com) and Powersim (http://www.powersim.com). The models presented here can therefore also be easily processed with these (and other) methods. All simulation models were tested in detail, in particular also with respect to the units of measurement (dimensions) used. In some models standardized nondimensional state variables normalized to "1" were used which, however, can also be easily converted to real dimensional formulations (cf. Bossel 2007 *Systems and Models*).

With few exceptions, the same notation is used for all model documentations: *Variables* are indicated by italics; small capitals are used for (mostly constant) PARAMETERS. State variables are shown as boxes in the system diagrams. Units of measurement are shown in [square brackets] in the program listings (but are not part of the respective equation!).

The software systems mentioned are particularly user-friendly. Their use can be learned quickly and easily. The software systems differ in small details, but operate in the same manner. To implement and compute the simulation models documented in the System Zoo the following steps have to be taken:

1.   *Enter simulation time parameters and save the model* under a name of its own. The time query usually appears in the first model frame; details can be changed later.
2.   *Place the system variables and parameters on the screen.* For this, select a corresponding button for (1) state variable, (2) rate of change, or (3) other system quantity, move the corresponding symbol to the desired place on the screen and place it there by mouse click. A rate (= "valve" symbol) is connected to a corresponding state variable by a "pipe". Enter the name of the variable or parameter.
3.   *Connect the different quantities by influence arrows.* Select the button for the influence arrows, click on the "sender" quantity, draw the arrow to the "receiver" quantity, and put the arrow down by mouse click. If the quantities are too far apart in the simulation diagram, they can be connected using "shadow" or "ghost" variables (shown in the diagrams in <pointed> brackets).
4.   *Quantify system quantities.* Click on the button for "equations", and then consecutively on each variable or parameter in the system diagram. A form appears with the name of the quantity (entered in Step 2) and the names of all input quantities connected to it (defined by the influence arrows as entered in Step 3). The mathematical function to be used for computing the quantity from the input quantities is entered in the form. For constant parameters the numerical values have to be entered.
5.   *Start the simulation.* The program system checks completeness and correctness of the model formulation and reports possible (formal) errors. If these have been corrected, the simulation can be started using the "run" button. Euler-Cauchy integration is usually used but the more exact Runge-Kutta method (RK4) can also be chosen.
6.   *Select the results and their presentation.* Every system quantity can be documented individually or together with other quantities using a variety of possible representations (diagrams, tables), in particular time diagrams or state space diagrams.

Simulation models of dynamic systems are mathematical models employing difference or differential equations which describe the (temporal) change of "state variables". It is not necessary to be familiar with this mathematical apparatus to work with simulation models and even to develop them. Working with the models of the System Zoo and learning from the methods employed there will provide valuable experience for developing your own models. Information on the theoretical and mathematical background of modeling and simulation of dynamic systems is found in an accompa-nying book (Bossel 2007 *Systems and Models*). This text presents elementary concepts like state equations, standardized and dimensionless quantities, equilibrium, oscilla-tion, stability and instability, linearization, limit cycle, chaos etc. which are necessary for more intensive work and a deeper understanding of a system's dynamics.

Notes on working with the models: Although the model descriptions differ, every documentation follows the same sequence: Description of modeling task, simulation model, and major structural characteristics of the system; complete simulation dia-gram (sometimes several diagrams); complete listing of model equations; description of representative reference runs with time diagrams and state diagrams; hints at un-usual features; exercises; references. Additional information on most models can be readily found in the internet.

All models are provided with default parameter settings which already demon-strate certain characteristic peculiarities. Suggestions for additional interesting investi-gations are provided with each model documentation. In addition to these model-specific suggestions the following general suggestions apply to all models:

1. *Start by examining the behavior of the reference run* (with the given default pa-rameter settings) using the different modes of representation of results provided by the simulation software (e.g. time diagrams, state space representation, tables).

2. *Investigate the dependence of system development on "critical" parameters* mentioned in the documentation. It is recommended to make several runs with differ-ent values for a particular parameter, save the results, and compare these using dia-grams or tables.

3. *Analyze the system in greater detail in parameter ranges where significant changes of system behavior can be observed* (e.g. stability/instability, equilib-rium/collapse) or where other interesting effects appear.

4. *Investigate the global behavior* (for systems with two state variables) in the com-plete (relevant) state space for the reference case and/or interesting combinations of parameters (a supplementary module for the generation of state space diagrams is provided by model Z115 "State space diagram"). In particular, find the equilibrium points and determine stability or instability from the state trajectories.

5. *Calculate (analytically) the location of equilibrium points* as function of the pa-rameters using the state equations and the condition that rates of change must disap-pear at the equilibrium points ($d\mathbf{z}/dt = \mathbf{0}$). Compare this with simulation results for the same parameter choice. (Cf. Bossel 2007 *Systems and Models* Ch. 2.8).

6. *Linearize the nonlinear state equations at the equilibrium points* and analyze the behavior of the corresponding linearized substitute system using the model of the lin-ear oscillator (of same order; see models Z114, Z117) by substituting corresponding system parameters. Does the behavior of the original nonlinear system near the equi-librium points agree with the properties of the substitute linearized system and its ei-genvalues? (cf. Bossel *Systems and Models* Ch. 2.8 for the theoretical background). (This suggestion refers primarily to two-dimensional systems and is intended for read-ers with some mathematical interest and background).

7. *Translate the (mostly) nondimensional generic models into simulation models for real systems* by correct dimensionalization of parameters and variables, and choice of suitable initial states and parameters. Compare the simulation results with experience and observations.

3

Climate and Vegetation

While mathematical formulae and calculations have always been an inseparable part of physics and engineering, applications of mathematics and computer simulation have only spread slowly to other fields of science. This is primarily due to the complexity of the systems which, for example, the biological, environmental and social sciences have to deal with. Often, they can only be made computable by heroic simplifications and brave omissions. In this way, seemingly unimportant components may be overlooked which are later found to have a decisive influence on behavior. The "butterfly effect" of chaotic systems must be a warning, but fortunately only a small number of the systems found in the real world prove to be chaotic.

The laws of nature require that in all areas of reality material and energy flows must strictly balance. Materials and energy cannot arise out of nothing or simply disappear again. Mathematical description of material and energy flows (using the conservation laws for materials and energy) can therefore often serve as reliable base for system models and computer simulation of processes in even very complex systems.

In this part of the System Zoo 14 simulation models are introduced whose structures are primarily determined by processes of material and/or energy transformations and conservation laws. In this manner reliable assessments for decision support and long term planning can often be obtained without minute representation of details. *Examples*: the calculation of stream levels in a water catchment basin after a heavy rain; the change of the global CO_2 balance by fossil fuel emissions and forest clearing; the influence of pollutants on forest growth; and optimal fertilization and irrigation in farming.

Z301 Regional water balance. If one imagines an envelope surface surrounding a water catchment area, the same amount of water must leave this surface by outflows and evaporation in the course of several years as has entered it through precipitation. However, the dynamics of outflows and evaporation is determined by many processes and different reservoirs. Water from precipitation partly evaporates on the ground and on vegetation, seeps away, is stored in the upper layers of the soil, seeps into groundwater reservoirs, leaves the ground in springs, flows away in creeks and rivers, or is stored in lakes and reservoirs. These processes depend on the characteristics of soils, vegetation, geology and orography. Many parameters and processes must be taken into account to calculate e.g. the temporal development of water flow and water level in a river after a heavy rain.

Z302 Global carbon cycle. The carbon dioxide level of the atmosphere remained almost unchanged for hundreds of thousands of years until the beginning of industrialization. This is evidence that carbon flows into and out of the atmosphere were in balance. The enormous amount of CO_2 taken up by plants every year corresponded exactly to the amount of CO_2 which, every year, reached the atmosphere by decomposition processes of organic matter. The global carbon dynamics was in balance. Since the beginning of industrialization, however, the atmospheric CO_2 level rises strongly – it has meanwhile increased by approximately 40 percent. The major causes are the burning of fossil fuels and the clearing of forests. In comparison with natural CO_2

processes anthropogenic contributions are small, but they suffice to completely upset the delicate atmospheric carbon balance. Since CO_2 is a greenhouse gas, this increase must be recognized as a major cause for higher average temperatures and climate change.

Z303 CO_2 dynamics of biosphere and atmosphere. A more detailed breakdown of the global carbon flows substantiates the conclusion that the CO_2 balance has been upset by human activities, and that the CO_2 level of the atmosphere will rise still further even if drastic measures are taken to control it. A simulation model of this type can help to assess more exactly the long-term consequences of suggested measures (e.g. increase of energy efficiencies, or reforestation) and to show the consequences of an unchecked increase of fossil energy consumption and deforestation.

Z304 Forest destruction and CO_2 dynamics. The destruction of tropical forests to create fields for agricultural production changes the CO_2 dynamics for several reasons. In agricultural areas, the storage capacity for carbon in plants, straw and soil humus is much less than in forests; it is even lower in degraded fallow areas. The transformation of forests to fields, and later of fields to degraded land releases additional amounts of carbon. Reforestation can partly make up for the losses and lead to sequestration of atmospheric CO_2. A simulation model allows calculation of the consequences and dynamics of these coupled processes and can contribute to better preparation of measures which are effective in the long run.

Z305 Carbon balance of forests. A more precise analysis of the role of forests in the global CO_2 balance requires a more detailed representation of carbon absorbing and carbon releasing processes in forests and in the utilization of wood. Initially carbon is assimilated from the atmosphere by the leaf mass, thus fixing solar energy in glucose. A large portion of this energy is consumed in respiration (i.e. for maintenance of life processes) of the plants; a smaller portion is stored in wood mass. Forest litter (leaves, deadwood) is decomposed and partly converted into humus. Trees are felled and used as firewood and timber. The carbon contained in timber returns to the atmosphere much later. A systems study allows drawing conclusions, for example about the long-term storage of carbon in forests as a function of mean annual temperature, or about differences between forests in tropical and temperate climates.

Z306 Motor vehicle traffic and CO_2 emissions. A large and growing share of global CO_2 emissions originates from motor vehicles. In many developing countries the number of motor vehicles and corresponding traffic volume and fuel consumption increase drastically because of population increase and strong economic growth. However, total fuel consumption and carbon emissions depend on the specific fuel consumption of motor vehicles. Efforts to reduce fuel consumption of new vehicles can therefore have considerable influence on magnitude and temporal development of CO_2 emissions. These relationships can be easily represented in, and computed with a dynamic simulation model. This allows investigation of scenarios using different assumptions about growth of the vehicle fleet and improvement of energy efficiency.

Z307 Photosynthesis of plants. The correct computation of the photoproduction of plants has to take into account a variety of factors which are partly a function of time of day, season and geographical latitude: temperature, position of the sun, clouds, duration of solar radiation. Depending on constantly changing radiation and temperature, plant-specific photosynthesis properties, leaf density, light attenuation in the plant canopy, and leaf respiration (for maintenance), the leaf canopy assimilates a time-variable amount of energy which can be expressed by its CO_2 assimilation. The daily production is the sum (the time integral) of instantaneous production; it is highest in summer. Such a dynamic simulation model of photoproduction is at the core of more complicated forest models which allow reliable computation of forest stand development under different management conditions over decades and centuries.

Z308 Forest dynamics. The growth dynamics of a forest stand arises from the energy surplus of photoproduction which remains for wood increment after deduction of energy consumed for maintenance and renewal of leaves, branches, stems and roots. At first the canopy will fill quickly to reach maximum energy assimilation. Self-shading of leaf layers restricts leaf density, however. A permanently growing structure must be maintained by the constant production of the canopy, so that wood increment declines gradually and finally disappears if energy gains are just compensated by energy losses. This mechanism prevents trees from growing indefinitely. An interesting dynamics results if (e.g. by air pollutants) the photosynthesis performance of the leaves is reduced. If a critical damage value is exceeded, then a very sudden collapse of the forest stand ("forest dieback") follows since the trees "starve" because of the negative energy balance.

Z309 Tree dieback. Air pollutants can impair the photosynthesis of leaves and reduce the assimilation of solar energy by photosynthates (glucose). If there is a photosynthate deficiency, leaves and feeder roots cannot be renewed in sufficient quantity to make up for annual losses. The supply situation worsens, water and nutrients cannot be taken up any more in sufficient quantity, and the tree can no longer accumulate enough energy for maintenance and finally dies. The process affects every part of the tree system with its interdependent processes and components. However, air pollutants can also lead to accelerated mortality of feeder roots by soil acidification, without photosynthesis being impaired directly. In this case, water and nutrient deficiencies lead to the same vicious cycle: The tree dies eventually (and rather suddenly) after a longer period of insufficient supply. These system relationships and their consequences become very obvious in a system model. It is a general characteristic of such "systemic diseases" that symptoms often do not provide any direct clues to causes which may be hidden in unexpected places.

Z310 Soil water dynamics. Simulation models of plant growth, which can be used to investigate consequences of different methods of cultivation, irrigation, fertilization, and pest and weed control for harvest yield and operating costs, play an important role in agricultural management. A detailed simulation model of the dynamics of plant-available soil water as function of soil and cultivation parameters, and as a result of precipitation, irrigation, percolation, evaporation, and transpiration of the plants is an important component of such a management tool. The dynamics of soil water and plant growth are intimately coupled – each causes and depends on the other. Field capacity of the soil, capillary rise dependent on soil type, and water holding capacity of soil organic matter play important roles in soil water supply. Soil water dynamics and plant growth are significantly affected by random precipitation patterns.

Z311 Nutrient dynamics. Plants can usually take up required nutrients from the soil where they become available by rock weathering and plant litter decomposition. To achieve high yields, some nutrients must be supplied in relatively large quantities as fertilizers if they are not available in sufficient quantity in the soil: nitrogen, phosphate, potash, lime and magnesium. While the other substances have "slow" dynamics and can act as fertilizer for several years, nitrogen fertilization is characterized by very "fast" dynamics. If nitrogen fertilization is not in harmony with plant growth, fertilizer and harvest losses, and environmental damage to atmosphere and water supply can be substantial. Since nitrogen can be fixed by soil bacteria and by leguminous plants, and can also be supplied with animal manure, organic farming operates without artificial nitrogen fertilizer. To obtain high yields despite this handicap, the different processes of nitrogen chemistry in soil, plant, and litter must be understood and utilized in the best possible way. However, these processes are primarily connected with the decomposition of organic material from animal manure, harvest waste and compost into nutrient humus and permanent humus. The model therefore deals with the coupled carbon and nitrogen transformations which provide the plant-available nitrogen which in turn determines plant growth and harvest yield.

Z312 Field crop cultivation. A simulation model of plant cultivation can only be a reliable planning aid to the farmer if it correctly represents not only the complex system of interacting processes of water and nutrient dynamics in the soil and in the plant but can also be adapted to local conditions using plant and soil specific parameters and real weather data. To accomplish this, the submodels for plant growth and for soil water, carbon, and nitrogen processes that were separately developed and checked are coupled to each other. The precipitation pattern can be matched to real weather conditions using a randomizer. Soil conditions can be accurately parameterized using several soil parameters. Plant-specific parameters can be applied for a variety of field crops from potatoes to wheat to calculate growth dynamics and harvest yields correctly. Application dates and amounts of organic and mineral fertilizers can be chosen to develop optimal fertilization strategies. The computed time paths of plant growth, nitrogen and water availability allow instructive insights into the dynamic processes taking place in the system, which would otherwise become evident only by constant and costly field measurements.

Z313 Food production. Humankind faces the task of providing a growing world population with sufficient food. Since the area for agricultural production can hardly be expanded, the problem seems soluble only by increasing harvest yields – from which is often derived the alleged necessity for intensive fertilization, pesticide use and genetically modified plants and animals. This overlooks the fact that the human menu consists of vegetable and animal food, and that for the production of an animal food energy unit about ten vegetable food energy units are required. The same agricultural area can therefore feed ten times as many people on a vegetarian diet as on a diet based exclusively on animal products. For example, if the diet habits shift from a share of animal food of 40% in industrial countries to a (far healthier) share of only 10%, then large amounts of grain would become available for the human diet. Using the model it is possible to investigate the enormous maneuvering space resulting from different scenarios of changing diet composition.

Z314 Agriculture and farm bankruptcy. Agricultural enterprises can only continue to exist if economic conditions permit this. Overproduction (partly subsidized) in the countries of the European Union has kept prices of agricultural products at a low level while the costs of agricultural production have increased considerably. Smaller farms are the first to be forced into bankruptcy since larger farms can work more cost-efficiently. As farms disappear, the social and settlement structures of whole areas change. The model takes into account – among other factors – the different contributions of state interventions to a farm's balance sheets and determines from net farming income the tendency toward either productivity increase or termination of the farming operation.

Z301 Regional water balance

Simulation task

All organisms need water to live and thrive. Local water availability determines the development opportunities of ecosystems, farms and settlements. Many factors and processes influence local water availability and its development in time: irregular and seasonally varying precipitation; absorption capacity of soil and vegetation; storage in ground water, lakes, and reservoirs; evaporation; consumption and runoff. These interconnected processes lead to complex dynamics. It is important to know and to understand them well and in detail – for example, for reliable flood prediction or agricultural management

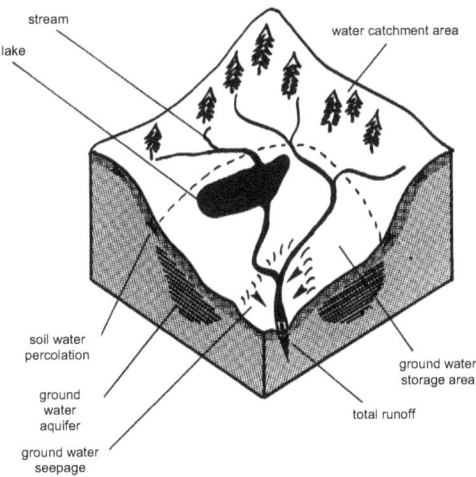

Figure Z301a: Water catchment area.

   The natural boundary for such investigations is that of the water catchment area. We consider an area separated by watersheds from other water catchment areas so that inflows consist exclusively of the precipitation over this area (Figure Z301a). Precipitation partly enters the soil and fills available storage capacity (field capacity) or flows away in surface runoff. Soil water partly seeps into the ground water, but is also partly evaporated on the surface (evaporation) or taken up by vegetation and transpired (transpiration). (Since both processes are difficult to separate in practice, their combined effect is usually considered as "evapotranspiration".) If the ground water level reaches the seepage level (i.e. on hillsides), it will flow to the surface in springs. Surface runoff and spring water may be partly captured by natural or artificial lakes and reservoirs, which serve to steady stream and river flow. The different storage volumes in the system (soil water field capacity, ground water aquifer, reservoirs) therefore have a smoothing effect on the flows in the system and serve to provide a steady and often fairly constant supply of water in the soil, the ground water aquifer, and in runoff. If storage volumes are large, the randomness of precipitation or even its seasonal variation will have little effect on these flows. However, if storage volumes are small, e.g. because of deforestation, removal of vegetation, or loss of organic matter in the soil (with its high water-holding capacity), this may lead to frequent and recurring floods on the one hand, and long periods of extreme drought on the other.

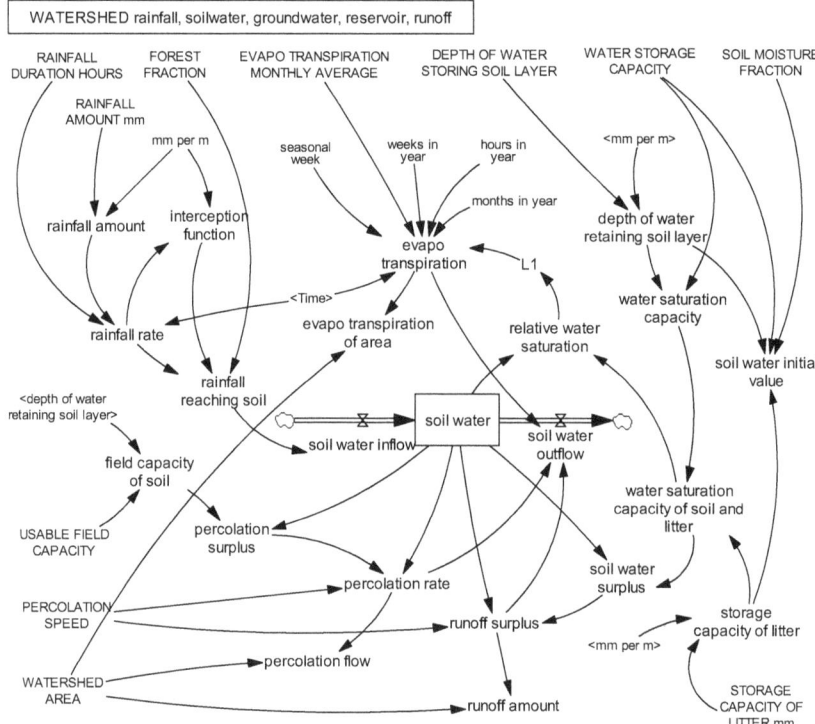

Figure Z301b: Simulation diagram of the regional water balance, part 1

Simulation model

The simulation diagram of the model is shown in Figures Z301b and c; the complete model equations are listed in the following. Depending on the DEPTH OF WATER STOR-ING SOIL LAYER, its WATER STORAGE CAPACITY and STORAGE CAPACITY OF LITTER, the soil can absorb precipitation as *soil water* as long as its *water saturation capacity* has not filled up. *Soil water inflow* depends on RAINFALL DURATION and RAINFALL AMOUNT, and on interception by vegetation, which is a function of FOREST FRACTION. Surplus *soil water* then percolates to lower levels, filling up the *ground water*. The *percolation rate* depends on soil parameters. If the soil is saturated, water remaining on the surface flows away as *runoff surplus*. The *relative water saturation* also deter-mines how much water evaporates on the surface and in vegetation. This *evapotran-spiration* itself is a (seasonal) function of insolation. The *ground water* supply is filled up by *percolation flow*. If the *height of ground water level* reaches the SEEPAGE HORI-

ZON, ground water drains out of the SEEPAGE AREA at the *seepage rate* determined by soil parameters. Since in general the *seepage rate* is limited, an overload of the ground water aquifer is possible, and the *runoff surplus* will flow off on the surface as *creek runoff*.

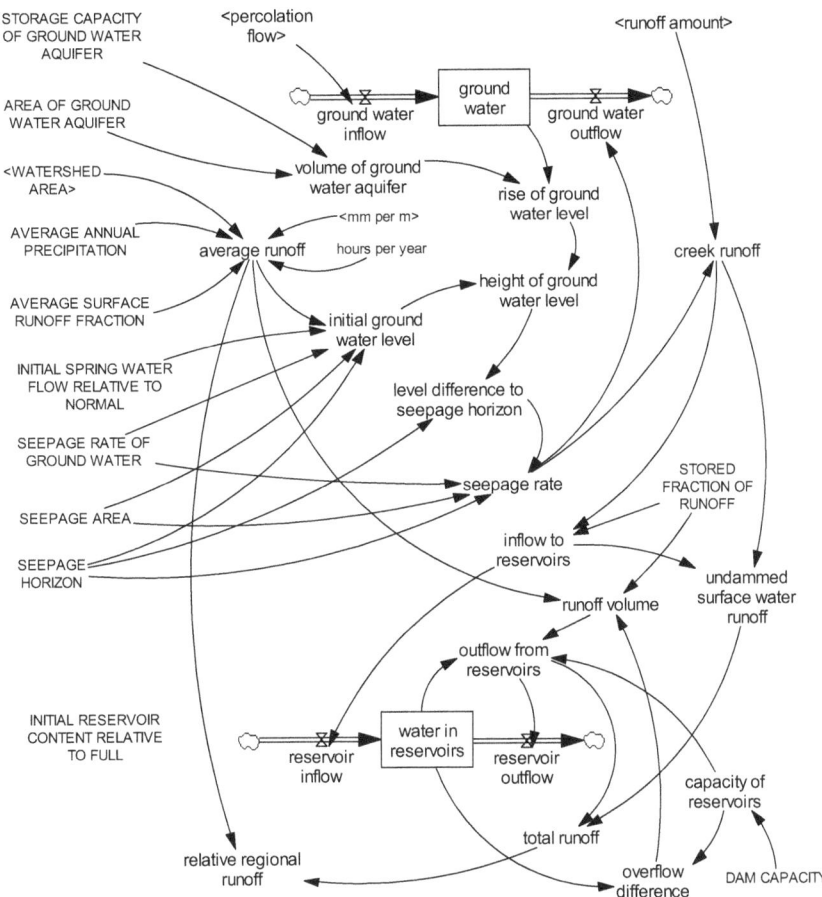

Figure Z301c: Simulation diagram of the regional water balance, part 2.

It is assumed here that part of the water flowing off superficially is accumulated in (natural or artificial) reservoirs or lakes. If the DAM CAPACITY is exceeded, overflow occurs. The *outflow from reservoirs* is a function of the volume of *water in reservoirs*, which regulates the outflow to some degree. The *total runoff* from the area is the sum of the *outflow from reservoirs* and the *undammed surface water runoff* resulting from surface runoff and ground water seepage. The *average runoff* can be used to determine the *relative regional runoff* indicating the steadiness of runoff.

The model employs a large number of parameters that allow adapting it to a variety of conditions. It is possible to adjust the WATERSHED AREA, the DEPTH OF WATER STORING SOIL LAYER, the STORAGE CAPACITY OF LITTER, the initial SOIL MOISTURE FRACTION, the DAM CAPACITY and the corresponding STORED FRACTION OF RUNOFF, the INITIAL RESERVOIR CONTENT RELATIVE TO FULL, the INITIAL SPRING WATER FLOW RELATIVE TO NORMAL, the initial SEASONAL WEEK of the simulation (important for the computation of seasonally varying evaporation), the RAINFALL DURATION HOURS, and the RAINFALL AMOUNT. Other parameters that can be easily changed by the user are: the USABLE FIELD CAPACITY of the soil, the PERCOLATION SPEED, the STORAGE CAPACITY OF GROUND WATER AQUIFER, the SEEPAGE HORIZON, the SEEPAGE RATE OF GROUND WATER, the SEEPAGE AREA, and the WATER STORAGE CAPACITY of soil. With these parameters very different conditions can be represented to examine their influence on the water balance of the water catchment area.

Regional parameters
WATERSHED AREA = 1e+008 [m*m]
FOREST FRACTION = 0 [1]
DAM CAPACITY = 2e+006 [m*m*m]
STORED FRACTION OF RUNOFF = 0.5 [1]
INITIAL RESERVOIR CONTENT RELATIVE TO FULL = 0.5 [1]

Soil parameters
SOIL MOISTURE FRACTION = 0.5 [1]
WATER STORAGE CAPACITY = 0.3 [m/m]
USABLE FIELD CAPACITY = 0.2 [m/m]
DEPTH OF WATER STORING SOIL LAYER = 30 [mm]
PERCOLATION SPEED = 0.0001 [m/Hour]
STORAGE CAPACITY OF LITTER mm = 1 [mm]

Groundwater and spring parameters
AREA OF GROUND WATER AQUIFER = 1e+007 [m*m]
STORAGE CAPACITY OF GROUND WATER AQUIFER = 0.15 [m/m, water fraction]
AVERAGE SURFACE RUNOFF FRACTION = 0.2 [1]
SEEPAGE AREA = 1e+006 [m*m]
SEEPAGE HORIZON = -10 [m]
SEEPAGE RATE OF GROUND WATER = 0.01 [m/Hour]
INITIAL SPRING WATER FLOW RELATIVE TO NORMAL = 1 [1]

Precipitation parameters
AVERAGE ANNUAL PRECIPITATION = 876 [mm/Year]
EVAPO TRANSPIRATION MONTHLY AVERAGE = 37/1000 [m/Month]
seasonal week = 25 [Week]
RAINFALL DURATION HOURS = 30 [Hour]
RAINFALL AMOUNT mm = 50 [mm]

Conversions
mm per m = 1000 [mm/m]
months in year = 12 [Month]
weeks in year = 52 [Week]
hours in year = 8760 [Hour]
hours per year = 8760 [Hour/Year]

Soilwater and evaporation

depth of water retaining soil layer = DEPTH OF WATER STORING SOIL LAYER
/mm per m [m]

field capacity of soil = USABLE FIELD CAPACITY *depth of water retaining soil layer
[m]

storage capacity of litter = STORAGE CAPACITY OF LITTER mm /mm per m [m]

water saturation capacity = depth of water retaining soil layer *WATER STORAGE
CAPACITY [m]

water saturation capacity of soil and litter = storage capacity of litter +water saturation
capacity [m]

rainfall amount = RAINFALL AMOUNT mm /mm per m [m]

rainfall rate = IF THEN ELSE(Time >= 0 :AND: Time < RAINFALL DURATION
HOURS, rainfall amount /RAINFALL DURATION HOURS, 0) [m/Hour]

interception function = WITH LOOKUP (rainfall rate *mm per m, ([(0,0) -(20,1)], (0,1),
(0.5,0.5), (10,0.05), (20,0.02))) [1]

rainfall reaching soil = rainfall rate *(1 -FOREST FRACTION *interception function)
[m/Hour]

relative water saturation = soil water /water saturation capacity of soil and litter [1]

L1 = IF THEN ELSE (relative water saturation > 1, 1, relative water saturation) [1]

evapo transpiration = L1 *EVAPO TRANSPIRATION MONTHLY AVERAGE
*(1 +(35/37) *SIN (6.28*((Time /hours in year) +(Seasonal week /weeks in year) -
0.25))) *months in year /hours in year [m/Hour]

evapo transpiration of area = evapo transpiration *WATERSHED AREA [m*m*m/Hour]

soil water initial value = (depth of water retaining soil layer *WATER STORAGE
CAPACITY +storage capacity of litter) *SOIL MOISTURE FRACTION [m]

soil water inflow = rainfall reaching soil [m/Hour]

soil water = INTEG (+soil water inflow -soil water outflow, soil water initial value) [m]

soil water surplus = soil water -water saturation capacity of soil and litter [m]

runoff surplus = IF THEN ELSE (soil water surplus > 0, (soil water surplus /soil water)
*100 *PERCOLATION SPEED, 0) [m/Hour]
*Assumption: runoff rate= 100*percolation rate*

soil water outflow = evapo transpiration +runoff surplus +percolation rate [m/Hour]

runoff amount = runoff surplus *WATERSHED AREA [m*m*m/Hour]

Groundwater, percolation, seepage

volume of ground water aquifer = AREA OF GROUND WATER AQUIFER *STORAGE
CAPACITY OF GROUND WATER AQUIFER [m*m]

initial ground water level = (-SEEPAGE HORIZON) *((average runoff *INITIAL
SPRING WATER FLOW RELATIVE TO NORMAL /(SEEPAGE RATE OF
GROUND WATER *SEEPAGE AREA)) -1) [m]

percolation surplus = soil water -field capacity of soil [m]

percolation rate = IF THEN ELSE (percolation surplus > 0, (percolation surplus
/soil water) *PERCOLATION SPEED, 0) [m/Hour]

percolation flow = WATERSHED AREA *percolation rate [m*m*m/Hour]

ground water inflow = percolation flow [m*m*m/Hour]

ground water outflow = seepage rate [m*m*m/Hour]

ground water = INTEG (+ground water inflow -ground water outflow,0) [m*m*m]

rise of ground water level = ground water /volume of ground water aquifer [m]

height of ground water level = initial ground water level +rise of ground water level [m]

level difference to seepage horizon = height of ground water level -SEEPAGE
HORIZON [m]

seepage rate = IF THEN ELSE(level difference to seepage horizon > 0,
    (level difference to seepage horizon /(-SEEPAGE HORIZON)) *SEEPAGE RATE
    OF GROUND WATER *SEEPAGE AREA, 0) [m*m*m/Hour]

Dams and runoff
capacity of reservoirs = DAM CAPACITY [m*m*m]
creek runoff = runoff amount +seepage rate [m*m*m/Hour]
inflow to reservoirs = STORED FRACTION OF RUNOFF *creek runoff [m*m*m/Hour]
outflow from reservoirs = IF THEN ELSE (capacity of reservoirs > 0, (runoff volume
    *water in reservoirs /capacity of reservoirs), 0) [m*m*m/Hour]
reservoir inflow = inflow to reservoirs [m*m*m/Hour]
reservoir outflow = outflow from reservoirs [m*m*m/Hour]
water in reservoirs = INTEG (+reservoir inflow -reservoir outflow, capacity of
    reservoirs *INITIAL RESERVOIR CONTENT RELATIVE TO FULL) [m*m*m]
undammed surface water runoff = creek runoff -inflow to reservoirs [m*m*m/Hour]
overflow difference = water in reservoirs -capacity of reservoirs [m*m*m]
average runoff = ((AVERAGE ANNUAL PRECIPITATION /hours per year) /mm per m)
    *(AVERAGE SURFACE RUNOFF FRACTION) *WATERSHED AREA
    [m*m*m/Hour]
runoff volume = IF THEN ELSE (overflow difference > 0, STORED FRACTION OF
    RUNOFF *average runoff *20, STORED FRACTION OF RUNOFF *average run-
    off *2) [m*m*m/Hour]
total runoff = undammed surface water runoff +outflow from reservoirs [m*m*m/Hour]
relative regional runoff = total runoff /average runoff [1]

Simulation time parameters
INITIAL TIME = 0 [Hour]
FINAL TIME = 100 [Hour]
TIME STEP = 0.03125 [Hour]

Simulation results

We compare two simulation runs here which differ only in the assumptions about
FOREST FRACTION, the DEPTH OF WATER STORING SOIL LAYER and the STORAGE CA-
PACITY OF LITTER. In both cases the simulation describes developments over 100
hours. Within the first 30 hours relatively strong rainfall is assumed with altogether 50
mm of rain. Within the remaining 70 hours until the end of the simulation period no
further precipitation occurs. Figure Z301d shows the results for a wooded area (FOR-
EST FRACTION = 1) and corresponding DEPTH OF WATER STORING SOIL LAYER of 300
mm and STORAGE CAPACITY OF LITTER of 10 mm of water. In the second simulation
run of Figure Z301e for a deforested area (FOREST FRACTION = 0) the DEPTH OF WA-
TER STORING SOIL LAYER was reduced to 30 mm and only a very small STORAGE CA-
PACITY OF LITTER (1 mm of water) was assumed.

    Interesting differences appear between the two simulation runs. While the deep
forest soil gradually fills up with water up to its field capacity, and then slowly loses
water by evaporation and percolation, only little water remains in the thin soil cover of
the deforested area. After the rain period it is again at the lower limit of its field capacity (the wilting point). The deep, water-storing soil of the wooded area leads to only
gradual change of the ground water level. Thanks to the greater storage capacity of the
soil and the much smaller surface runoff the dams and lakes are not filled to capacity,

overflow does not occur, and the reservoirs therefore contribute to steadying the total runoff. If the soil cannot store much water, much more surface runoff occurs, reservoirs overflow, and flood peaks appear in the runoff.

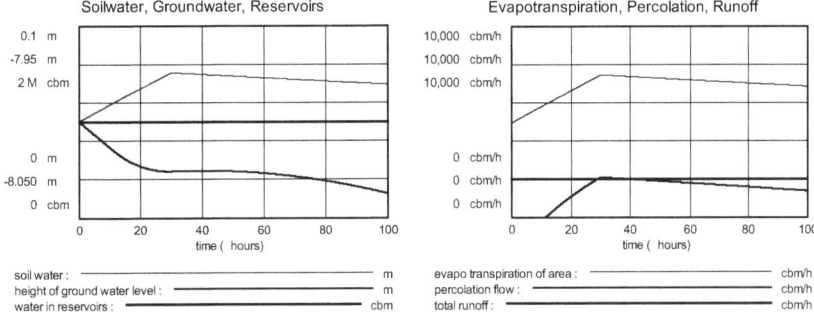

Figure Z301d: In a forest area with corresponding high water-holding capacity of the soil the total runoff hardly changes even after heavy rain.

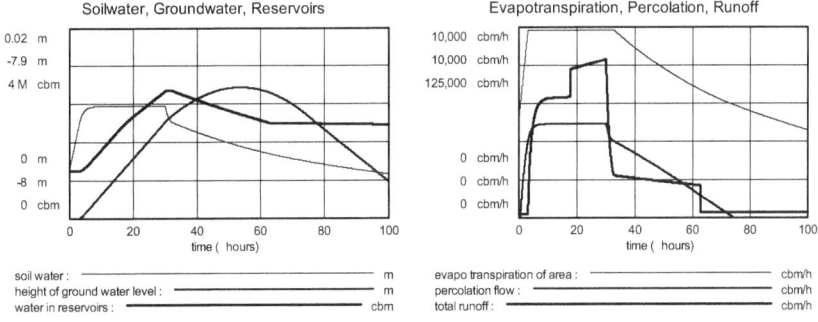

Figure Z301e The total runoff after heavy rain increases enormously if the catchment area lacks forest cover and soil water-holding capacity is low.

Exercises

1. Examine the negative effects of deforestation and soil erosion on the steadiness of water flow and availability. How much can they be reduced by dams and reservoirs?

2. Make the model more realistic by adding a stochastic calculation of rainfall based on climatological data.

3. Apply the model to a particular water catchment area (i.e. mountain valley, river basin) using realistic data. Investigate the consequences of measures like forest clearing, forest dieback, drainage, channeling of rivers and creeks, reduction of soil humus fraction, erosion, dams, ground water withdrawal by industry and agriculture, sealing of ground surface by roads and buildings etc. on the steadiness of water flow and availability (supplement the model if necessary).

4. Couple the model to a vegetation model for more exact computation of water consumption during the vegetation period (higher transpiration, lower evaporation) for different kinds of vegetation. Introduce a "sealing" of the ground by frost and snow for the winter season. Simulate the seasonal change of water storage and water runoff.

References

Council on Environmental Quality: *Global 2000 – Der Bericht an den Präsidenten.* Zweitausendeins, Frankfurt/M. 1980, S. 347-385, S. 701-706.
Larcher, W. 1979: *Ökologie der Pflanzen.* UTB/Ulmer Stuttgart, S. 281-357.
Finck, A. 1982: *Pflanzenernährung in Stichworten.* Hirt, Kiel, S. 75-81.
Scheffer, F., Schachtschabel 1982: *Lehrbuch der Bodenkunde.* Enke, Stuttgart, S.161-189.

Z302 Global carbon cycle

Simulation task

By photosynthesis and decomposition of organic matter (stand litter and humus) and by respiration of plants and animals large amounts of carbon dioxide are constantly being removed from and returned to the atmosphere. These gigantic CO_2 flows were in equilibrium over millions of years. Annual CO_2 gains and losses of the atmosphere balanced rather exactly, so that the atmospheric CO_2 level hardly changed.

Since the beginning of industrialization this dynamic equilibrium between the CO_2 reservoirs of atmosphere and (living and dead) biomass has been disturbed by the burning of fossil fuels and the deforestation of large areas. Every year more CO_2 now reaches the atmosphere than is taken out by photosynthesis. This leads to an increasing fraction of the greenhouse gas CO_2 in the atmosphere – a major cause of gradual temperature increase and of climate change.

The fact that the additional CO_2 input from human activities amounts to only few per cent of the natural CO_2 flows in the atmosphere leads to the often-heard assertion that this "small" effect cannot possible be a cause of climate change. Such assertions ignore the fundamental fact that even small perturbations of exactly balanced flow equilibriums always lead to dynamic and often dramatic changes in systems.

The phenomenon of dynamic equilibrium of flows between two reservoirs or pools is found in different areas. Two pools with contents x and y of a substance are in mutual exchange. A first process takes the substance at a certain rate from pool x to pool y. A second process in turn takes the substance from pool y and returns it to pool x. Flow equilibrium exists as long as no losses to or inputs from the external environment occur. This process applies generally to the dynamics of processes such as greenhouse gases in the atmosphere, absorption and decomposition of substances in the soil, in ground water, and in surface waters, rising pollution levels because of insufficient absorption, population development, financial balance sheets, and many technical processes.

Simulation model

The simulation diagram of Figure Z302a as well as the following model equations document the simple model of global carbon flows.

The two pools *carbon in atmosphere* and *carbon in biosphere* are linked to each other by the processes of *absorption by plants* and *decomposition of biomass*. This (initially balanced) carbon cycle is perturbed by the additional CO_2 input from burning of fossil fuels (with a logistic saturation of consumption) and forest destruction. This simple model does not take into account (the relatively small) uptake of atmospheric CO_2 by oceans.

Essential and critical factors for the further development of the system are the GROWTH RATE (of fossil fuel use) and the SATURATION VALUE FOSSIL EMISSIONS as well as the parameters of FOREST DESTRUCTION. Since the CO_2 uptake is independent of *carbon in atmosphere*, further CO_2 increase can only be avoided if the burning of fossil resources and forest destruction are stopped.

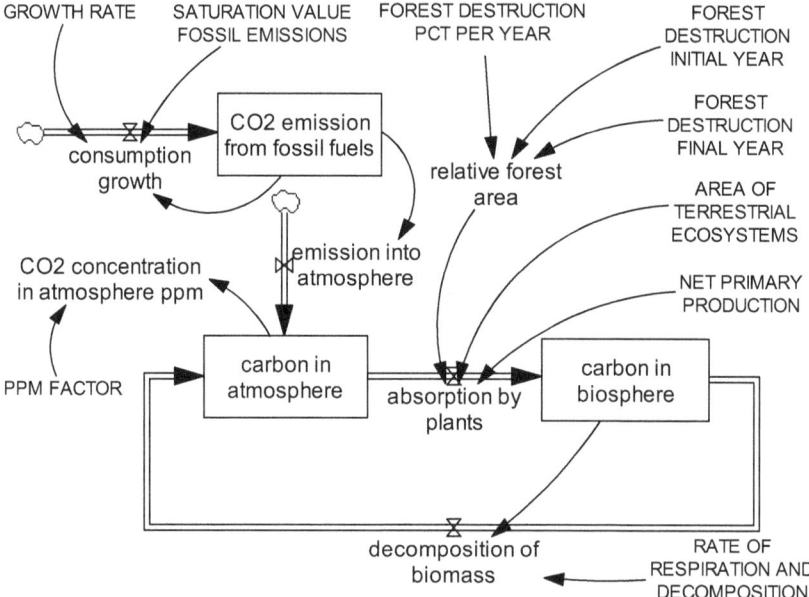

Z302a: Simulation diagram for the global carbon circulation.

Parameters
AREA OF TERRESTRIAL ECOSYSTEMS = 0.145 [Gkm²]
NET PRIMARY PRODUCTION = 400 [GtC/(Gkm²*Year)]
RATE OF RESPIRATION AND DECOMPOSITION = 0.02 [1/Year]
FOREST DESTRUCTION PCT PER YEAR = 0.2 [1/Year] pct per year
FOREST DESTRUCTION INITIAL YEAR = 1970 [Year]
FOREST DESTRUCTION FINAL YEAR = 2020 [Year]
SATURATION VALUE FOSSIL EMISSIONS = 15 [GtC/Year]
GROWTH RATE = 0.03 [1/Year]
PPM FACTOR = 2.12 [GtC/CO2ppm]

Dynamics
relative forest area = 1 –RAMP ((FOREST DESTRUCTION PCT PER YEAR /100),
    FOREST DESTRUCTION INITIAL YEAR, FOREST DESTRUCTION FINAL
    YEAR) [1]
consumption growth = GROWTH RATE *CO2 emission from fossil fuels *(1 -CO2
    emission from fossil fuels /SATURATION VALUE FOSSIL EMISSIONS)
    [GtC/(Year*Year)]
CO2 emission from fossil fuels = INTEG (consumption growth, 0.1) [GtC/Year]
emission into atmosphere = CO2 emission from fossil fuels [GtC/Year]
absorption by plants = NET PRIMARY PRODUCTION *AREA OF TERRESTRIAL
    ECOSYSTEMS *relative forest area [GtC/Year]

decomposition of biomass = RATE OF RESPIRATION AND DECOMPOSITION
    *carbon in biosphere [GtC/Year]
carbon in atmosphere = INTEG (+decomposition of biomass -absorption by plants
    +emission into atmosphere, 570) [GtC]
carbon in biosphere = INTEG (absorption by plants -decomposition of biomass, 2900)
    [GtC]
CO2 concentration in atmosphere = carbon in atmosphere /PPM FACTOR [CO2ppm]

Simulation time parameters
INITIAL TIME = 1850 [Year]
FINAL TIME = 2050 [Year]
TIME STEP = 0.05 [Year]

Simulation results

Figure Z302b shows the simulation result for the default parameter setting. This corresponds roughly to historical development. The *CO2 concentration in atmosphere* increases from a level of 280 ppm (parts per million) in year 1850 to a level of 650 ppm
in year 2050. Figure Z302c shows results for different SATURATION VALUE FOSSIL
EMISSIONS in the range from 5 to 25 GtC/yr (1 GtC/yr = 10^9 metric tons of carbon per
year). Here the *CO2 concentration in atmosphere* increases to values between 530 to
730 ppm in year 2050.

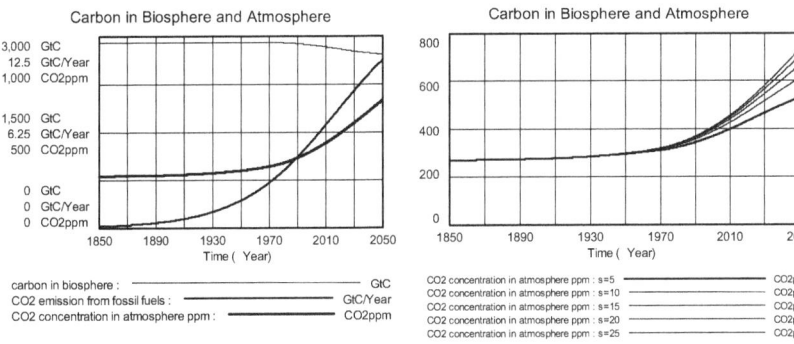

Figure Z302b: Historical and future development of the CO_2 level.
Figure Z302c: CO_2 level for different SATURATION VALUE FOSSIL EMISSIONS.

Starting out from initial equilibrium between the pools the level of *carbon in at-*
mosphere rises by *emission into atmosphere* caused by logistic growth of *CO₂ emis-*
sion from fossil fuels. The *absorption by plants* of CO_2 from the atmosphere depends
on the AREA OF TERRESTRIAL ECOSYSTEMS covered by vegetation and its mean NET
PRIMARY PRODUCTION (absorption of carbon by photosynthesis). FOREST DESTRUC-
TION or environmental pollution reduces the *absorption by plants* by the factor *rela-*
tive forest area. Unlike the process of photosynthesis the processes of decomposition
and respiration are proportional to the (living and dead) biomass, i.e. *carbon in bio-*
sphere with its corresponding RATE OF RESPIRATION AND DECOMPOSITION. Therefore

a significant input from respiration and decomposition arises even if biomass production should be reduced. Altogether, inputs to *carbon in atmosphere* then outweigh *absorption by plants,* leading to a steady increase of the *CO2 concentration in atmosphere.*

Exercises

1. Examine the further development for different scenarios of future consumption of fossil fuels (parameters: GROWTH RATE, SATURATION VALUE FOSSIL EMISSIONS).
2. Examine the consequences of different deforestation or afforestation strategies (FOREST DESTRUCTION PCT PER YEAR, FOREST DESTRUCTION INITIAL YEAR, FOREST DESTRUCTION FINAL YEAR).
3. Which role does the RATE OF RESPIRATION AND DECOMPOSITION play in the system? What does its reciprocal (the time constant) mean?
4. Using simulations develop suggestions for energy and forest policies that serve to stabilize the *CO2 concentration in atmosphere* at 400 ppm in the long run.

References

Bach, W. 1982: *Gefahr für unser Klima – Wege aus der CO_2-Bedrohung durch sinnvollen Energieeinsatz.* C. F. Müller, Karlsruhe (bes. S. 65-94).
Council on Environmental Quality, 1980: *Global 2000 – Der Bericht an den Präsidenten.* Zweitausendeins, Frankfurt/M, S. 548-574.
Ehrlich, P. R., Ehrlich, A. H., Holdren, J. P. 1977: *Ecoscience – Population, Resources, Environment.* W. H. Freeman, San Francisco.

Z303 CO$_2$ dynamics of biosphere and atmosphere

Simulation task

Carbon dioxide represents only a tiny portion of the earth's atmosphere with a share of about 0.038 percent (380 ppm). It nevertheless has a pivotal role for all life processes for two different reasons: First, carbon is the basic constituent of all organic life, with the earth's atmosphere serving as a reservoir of this vital element for all terrestrial and part of aquatic life. Second, the minute fraction of carbon dioxide in the atmosphere regulates to a significant degree the heat balance of the earth: Change of the carbon dioxide level causes global temperature change and corresponding climate change. Since the middle of the 19th century the CO$_2$ level has steadily increased as a result of burning of fossil fuels, intensification of agricultural production and deforestation, however.

Before the advent of significant perturbations by human interventions, the global carbon cycle was in stable flow equilibrium. The exchange between the large pools (atmosphere, living biomass, dead biomass (humus), oceans, and the earth's crust) was balanced. Every year, each pool released approximately the same amount of carbon as it gained during this period. The fixation of carbon in biomass by net primary production, and the reverse process by decomposition of organic matter represent the largest carbon flows by far between atmosphere and land surface. Additional natural carbon flows from land to atmosphere arise from animal respiration (heterotrophic respiration), from natural fires, and by rock weathering. In undisturbed state, the exchange between ocean and atmosphere is in balance; the net uptake of oceans or atmosphere is then zero.

Particularly since the 19th century, humans have increasingly intruded on this natural cycle, with the net effect that more carbon enters the atmosphere than can in turn be released to land or ocean. As a consequence the carbon dioxide level of the atmosphere has risen within the last 100 years from about 280 ppm to more than 380 ppm today – an increase of about 35 percent which continues to grow at a current rate of about 1.2 ppm per year (ppm = volume parts per million volume parts). This increase has several causes. A considerable share stems from growing use of fossil fuels. This value is relatively well known. Other contributions that are more difficult to determine arise from increasing deforestation particularly in tropical areas as well as from additional oxidation of humus by intensification of agriculture. The estimation of the individual contributions is difficult and involves significant uncertainties. Reference is made to the extensive literature on this subject.

The stocks and flow rates of the carbon cycle are expressed in (metric) gigatons of carbon (GtC, or GtC/yr). Significant uncertainties apply to most of the numbers. The atmosphere stores about 710 GtC today. This value is comparable to the estimated carbon content of living biomass of about 670 GtC. The estimated amount of carbon in dead organic matter of about 1600 GtC (fast decomposing nutrient humus and slowly decomposing permanent humus) is around two and a half times as large as the content in living biomass. The extractable amount of fossil fuels is estimated at about 3000 GtC. The most important carbon flows into or out of the atmosphere are the following today (in GtC/yr; gains positive, losses negative): rock weathering (0.5), burning of fossil fuels (5), fire (2), reforestation (primarily in northern latitudes) (-1), deforestation and forest clearing (2.5), of which a part enters the atmosphere directly by combustion (2.5), and another is initially fixed as charcoal in humus (1), animal respiration (3), net primary production of vegetation (-50), decomposition of dead biomass (45), additional soil oxidation by intensive agriculture (1.5), and net uptake of oceans (up to -3).

Climatology assumes today that for a doubling of the CO_2 content of the atmosphere from 300 to 600 ppm temperature averages over the northern hemisphere will increase by about 2 to 3 degrees Kelvin (centigrade). The increase would be particularly strong in polar latitudes at about 8 degrees, causing a melting of arctic ice masses and rising sea-level. Paradoxically, the former could disrupt the Gulf Stream, causing a new ice age in Northern Europe.

Global climate shifts are therefore to be expected, which could have serious consequences particularly for food security. For a quadrupling of the CO_2 content to 1200 ppm the summer drift ice of the northern and southern hemisphere would disappear completely, with drastic consequences for the global distribution of precipitation and corresponding effects on water supplies and agriculture.

The task is therefore to determine the possible future increase of the atmospheric CO_2 level, identify its major causes, and determine which measures must be taken to keep this increase and its consequences within limits. Since we are dealing with perturbations of the flow equilibrium of a dynamic system, a dynamic system model is in order. Particular attention will have to be given to the exploitation dynamics of fossil fuels because their utilization causes the largest portion of the CO_2 increase. It is particularly important to clarify whether and how the consumption of fossil fuels can be reduced by more efficient energy use and use of renewable energy, thus defusing the problem.

Simulation model

The structure of the model is represented in the simulation diagram of Figure Z303a; the corresponding model equations are documented in the following. The model calculates and keeps track of the carbon inputs and outputs of the atmosphere by rock weathering, fossil fuel use, fire, afforestation and deforestation, net primary production of vegetation, litter decomposition, accelerated humus decomposition by agriculture, and absorption of carbon dioxide in oceans. The consumption level of fossil fuels is controlled by their relative scarcity. Initially, consumption grows exponentially. It reaches a maximum and then reduces to zero when all exploitable resources have been used up. The initial values of the stocks are those for the year 1850. The simulation time period runs from 1850 to 2350.

The increase in *CO_2 in atmosphere* arises from the sum of *CO_2 increase* and *CO_2 decrease* taking into account *CO_2 uptake of oceans* which is represented simply by a table function. The maximum uptake is restricted to 3 GtC per year. The carbon content of the atmosphere is converted using the PPM FACTOR of 2.12 GtC per ppm carbon dioxide. The annual input from ROCK DECOMPOSITION is set to 0.5 GtC/yr. The input by natural fires (*loss from fires*) is assumed to be proportional to *net primary production* of plants NPP (Factor 0.035). The carbon uptake from the atmosphere by plants (*absorption by primary production*) by *net primary production* of plants is assumed to be proportional to *carbon in living biomass* (factor 0.075). The release of carbon to the atmosphere by animal respiration (*loss from animal respiration*) is proportional to NPP (factor 0.06). The *annual litter fall* corresponds to *net primary pro-*

duction (factor 0.905) (i.e. the amount remaining after plant respiration). The corresponding amount is deducted from *carbon in living biomass* and added by decomposition (*gain of carbon in humus*) to *carbon in humus*. Humification reduces this stock by *loss of carbon in humus*, which leads to *CO₂ increase* in the atmosphere. The input from *humus decomposition* is therefore proportional to *carbon in humus* (factor 0.032).

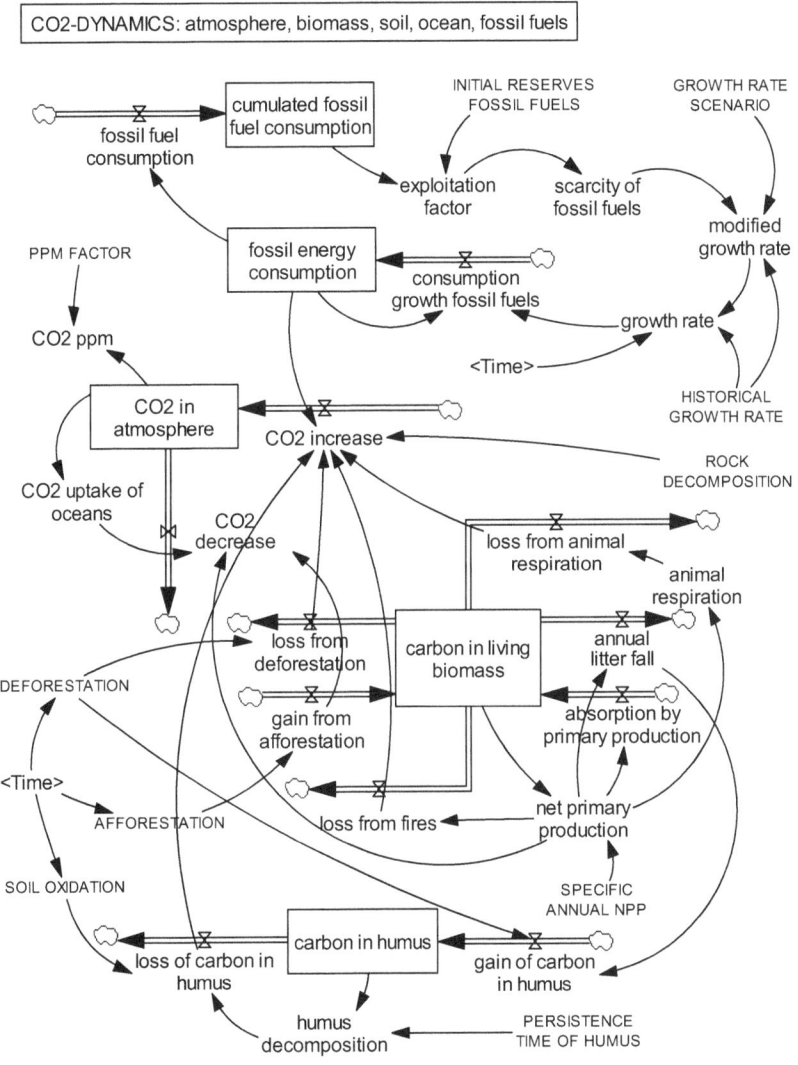

Figure Z303a: Simulation diagram for the CO₂ dynamics of biosphere and atmosphere.

For the carbon inputs and outputs resulting from AFFORESTATION, DEFORESTA-
TION and intensive agriculture (SOIL OXIDATION) corresponding time-dependent table
functions are provided. The DEFORESTATION data take into account that initially part
of the carbon released in forest clearing is fixed for some time as charcoal in humus.

The computation of the input from fossil fuel combustion assumes that the consumption level (*fossil energy consumption*) is a state variable which can change only
gradually. This change is initially determined by HISTORICAL GROWTH RATE or
GROWTH RATE SCENARIO which is modified however as soon as *scarcity of fossil fuels*
appears. This variable is computed by integrating the annual *fossil fuel consumption* to
obtain *cumulated fossil fuel consumption*. By comparing with INITIAL RESERVES FOS-
SIL FUELS an *exploitation factor* is computed. It assumes the value 1 when half of the
reserves have been used up. From this instant on the rate of increase of energy consumption (*consumption growth fossil fuels*) becomes negative as a result of *scarcity of
fossil fuels*. The *fossil energy consumption* finally decreases to zero if all reserves have
been used up. *Note:* This calculation approach only makes sense if the *cumulated fossil fuel consumption* at the beginning of the calculation is much smaller than INITIAL
RESERVES FOSSIL FUEL. Another formulation of this process – to which this restriction
does not apply – is found in model Z415 "Discovery of resources".)

Parameters
INITIAL RESERVES FOSSIL FUELS = 4000 [GtC]
HISTORICAL GROWTH RATE = 3.4 [1/Year]
GROWTH RATE SCENARIO = 3.4 [1/Year] *growth rate of fossil energy consumption
        beginning year 2000; pct per year*
DEFORESTATION = WITH LOOKUP (Time, ([(1850, 0) -(2300, 10)], (1850,0),
        (1980,2), (2100,1), (2400,1))) [GtC/Year]
AFFORESTATION = WITH LOOKUP (Time, ([(1850,0) -(2400,10)], (1850,0),
        (1950,1), (2100,1), (2400,1))) [GtC/Year]
ROCK DECOMPOSITION = 0.5 [GtC/Year]
PERSISTENCE TIME OF HUMUS = 30 [Year]
SOIL OXIDATION = WITH LOOKUP (Time, ([(1850,0) -(2350,10)], (1850,0),
        (1980,1.5), (2100,0), (2400,0))) [GtC/Year]
SPECIFIC ANNUAL NPP = 0.075 [1/Year]
PPM FACTOR = 2.12 [GtC/ppm]

Dynamics atmosphere
fossil fuel consumption = fossil energy consumption [GtC/Year]
cumulated fossil fuel consumption = INTEG (fossil fuel consumption, 10) [GtC]
exploitation factor = cumulated fossil fuel consumption /(INITIAL RESERVES FOSSIL
        FUELS/2) [1]
scarcity of fossil fuels = 1 -exploitation factor [1]
modified growth rate = IF THEN ELSE ((scarcity of fossil fuels *HISTORICAL
        GROWTH RATE < GROWTH RATE SCENARIO), (scarcity of fossil fuels
        *HISTORICAL GROWTH RATE), GROWTH RATE SCENARIO) [1/Year]
growth rate = IF THEN ELSE(Time<2000, HISTORICAL GROWTH RATE/100,
        modified growth rate/100) [1/Year]
consumption growth fossil fuels = growth rate *fossil energy consumption
        [GtC/(Year*Year)]
fossil energy consumption = INTEG (consumption growth fossil fuels, 0.1) [GtC/Year]

CO2 increase = fossil energy consumption +ROCK DECOMPOSITION +loss from
    fires +loss from deforestation +loss from animal respiration +loss of carbon in
    humus [GtC/Year]
CO2 uptake of oceans = WITH LOOKUP (CO2 in atmosphere, ([(0,0) -(6000,10)],
    (600,0), (700,3), (800,3), (5000,3))) [GtC/Year]
CO2 decrease = net primary production +gain from afforestation +CO2 uptake of
    oceans [GtC/Year]
CO2 in atmosphere = INTEG (CO2 increase -CO2 decrease, 600) [GtC]
CO2 ppm = CO2 in atmosphere /PPM FACTOR [ppm]

Dynamics biosphere
net primary production = SPECIFIC ANNUAL NPP *carbon in living biomass
    [GtC/Year]
absorption by primary production = net primary production [GtC/Year]
annual litter fall = 0.905 *net primary production [GtC/Year]
loss from deforestation = 0.6 *DEFORESTATION [GtC/Year]
loss from fires = 0.035 *net primary production [GtC/Year]
gain from afforestation = AFFORESTATION [GtC/Year]
animal respiration = 0.06 *net primary production [GtC/Year]
loss from animal respiration = animal respiration [GtC/Year]
carbon in living biomass = INTEG (absorption by primary production -annual litter fall
    -loss from deforestation -loss from fires -loss from animal respiration +gain from
    afforestation,750) [GtC]
gain of carbon in humus = annual litter fall +0.4*DEFORESTATION [GtC/Year]
humus decomposition = carbon in humus /PERSISTENCE TIME OF HUMUS
    [GtC/Year]
loss of carbon in humus = humus decomposition +SOIL OXIDATION [GtC/Year]
carbon in humus = INTEG (+gain of carbon in humus -loss of carbon in humus,1600)
    [GtC]

Simulation time parameters
INITIAL TIME = 1850 [Year]
FINAL TIME = 2350 [Year]
TIME STEP = 0.5 [Year]

Simulation results

For the model to be valid, its description of the historical development since the beginning of industrialization until today must be accurate. As a first step, the results
were therefore checked for the period from 1850 to 1975. They depend on uncertain
assumptions concerning afforestation, deforestation and clearing, and soil oxidation.
The values chosen on the basis of existing data represent developments with reasonable accuracy. It can therefore be assumed that the model is valid for estimates of
future developments.

    The time period from 1850 up to the year 2350 was used for further simulation
runs. Starting out from a current rate of increase of consumption of 3.4 percent per
year the annual *fossil energy consumption* climbs to a maximum around the year 2050,
decreasing to almost zero when reserves have been almost used up around the year
2200 (Figure Z303b). At this time the CO₂ level has reached a value of almost 2000
ppm. Only after exhaustion of the fossil fuel stocks can this high level be gradually
reduced again by the CO₂ uptake of the ocean. The effect of this reduction is relatively

insignificant since it proceeds only at a slow rate. These results agree approximately with more detailed calculations (cf. Bach 1982: 93-94).

Figure Z303c shows the development if the consumption of fossil energy would drop by 0.5 percent every year (GROWTH RATE SCENARIO) starting in year 2000. Even under this extreme assumption the CO_2 level would still increase to over 1000 ppm.

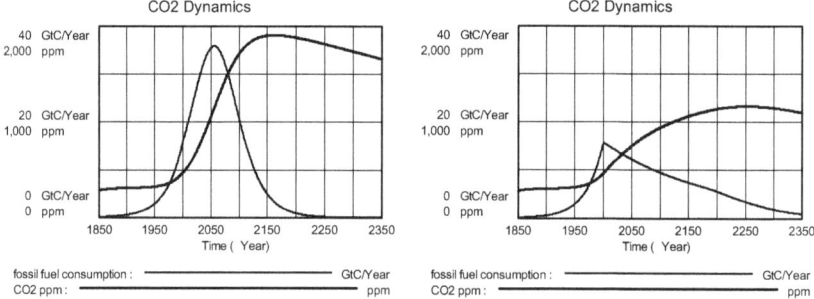

Figure Z303b: Development of the CO_2 level for a continuation of the historical growth rate of fossil energy consumption of 3.4 percent per year.
Figure Z303c: Development of the CO_2 level if fossil energy consumption decreases by 0.5 percent per year starting in year 2000.

Exercises

1. Implement the model and reproduce the results shown here for 1850 to 2350.
2. How does the CO_2 level increase for different assumptions for the GROWTH RATE SCENARIO of fossil energy consumption after the year 2000
    for the reference case (3.4 percent per year)?
    for higher growth (e.g. 6 percent)?
    for energy conservation (e.g. 2, 1, 0, -1 percent for 50 years, then 0 etc.)?
Which maximum CO_2 level is reached in each case?
3. Suggest an ecologically sound reforestation and timber harvesting policy and examine the achievable results (in combination with a sensible energy policy) for the next centuries (DEFORESTATION, AFFORESTATION, and GROWTH RATE SCENARIO for fossil energy consumption after 2000).
4. The present formulation of CO_2 uptake by the oceans is physically unsatisfactory (but simple) since the exchange equilibrium between atmosphere and ocean is not taken into account explicitly. Find a scientifically more satisfactory representation.

References

Bach, W. 1982: *Gefahr für unser Klima – Wege aus der CO₂-Bedrohung durch sinnvollen Energieeinsatz.* C.F. Müller, Karlsruhe, bes. S. 65-94.
Bach, W. u.a. 1980: The carbon dioxide problem - an interdisciplinary survey. *Experientia separatum,* vol. 36, FASC. 7, S. 767-890, Birkhäuser, Basel.
Colinvaux, P. 1993: *Ecology* 2. J. Wiley, New York (S. 589-618).
Ehrlich, P. R., Ehrlich, A. H., Holdren, J. P. 1977: *Ecoscience: Population, Resources, Environment.* W. H. Freeman, San Francisco (S. 67-95).

Z304 Forest destruction and CO_2 dynamics

Simulation task

Forests store more carbon in biomass and soil than fields under agricultural cultivation, and these in turn store more carbon than fallow land. If land use changes into another mode by clearing, afforestation, degradation, and housing development, corresponding amounts of CO_2 are released into the atmosphere or stored in biomass and humus. The CO_2 flows arising from changes of land use can be computed from the dynamics of change of land areas and the specific values of carbon storage in the different land use modes. Of particular interest are the CO_2 flows resulting from the clearing of tropical forests and their successive transformation to farmland, pastureland or degraded land.

Simulation model

The simulation model is documented completely in Figure Z304a and the following model equations. A constant TOTAL AREA is assumed which corresponds to the area covered originally by tropical forests. The state variables *forest area* and *agricultural area* – both changing with time – are deducted from the TOTAL AREA; the residual area is *degraded land area* (including settlements, streets, fallow land etc.).

CO_2 releases (or CO_2 sequestrations) result from (positive or negative) storage differences between original and subsequent land use modes and annual change of land use in the different categories by the processes *deforestation, reforestation and regeneration* as well as *degradation* (fallowing, housing developments, road construction etc.). For DEFORESTATION, REFORESTATION and DEGRADATION after 1990 appropriate scenario parameters can be provided to allow investigation of different developments and their consequences for CO_2 release.

Parameters
TOTAL AREA = 4e+007 [km²]
FOREST AREA 1990 = 1.8e+007 [km²]
AGRICULTURAL AREA 1990 = 2e+007 [km²]
DEFORESTATION 1990 = 250000 [km²/Year]
ANNUAL DEFORESTATION AFTER 1990 = 250000 [km²/Year]
REFORESTATION 1990 = 10000 [km²/Year]
ANNUAL REFORESTION AFTER 1990 = 10000 [km²/Year]
LAND DEGRADATION RATE 1990 = 0.005 [1/Year]
LAND DEGRADATION RATE AFTER 1990 = 0.005 [1/Year]
CO2 SEQUESTRATION IN FARMLAND = 25000 [tCO2/km²]
CO2 SEQUESTRATION IN FALLOW LAND = 5000 [tCO2/km²]
CO2 SEQUESTRATION IN FOREST LAND = 100000 [tCO2/km²]
RECOVER TIME OF DEGRADED SOIL = 200 [Year]

Dynamics
degraded land area = TOTAL AREA –forest area -agricultural area [km²]
regeneration rate = 1/RECOVER TIME OF DEGRADED SOIL [1/Year]
regeneration = degraded land area *regeneration rate [km²/Year]
reforestation = IF THEN ELSE (Time <= 1990, REFORESTATION 1990, ANNUAL
    REFORESTION AFTER 1990) [km²/Year]

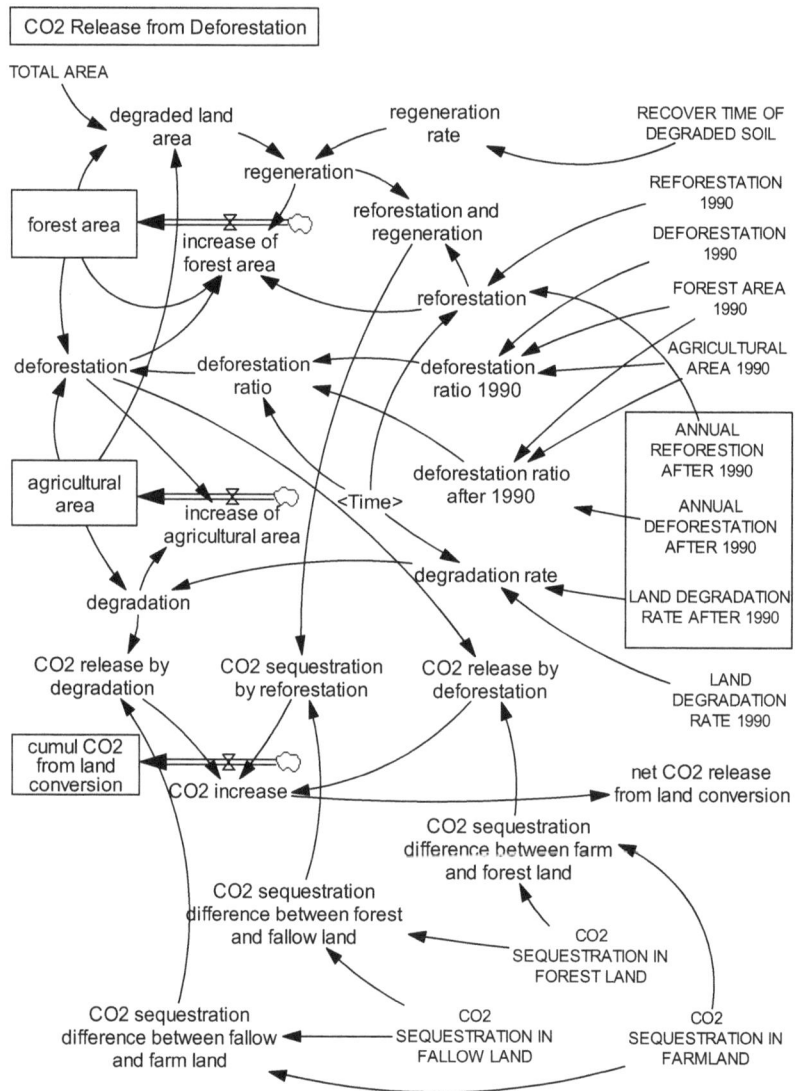

Figure Z304a: Simulation diagram for CO_2 dynamics from forest destruction.

reforestation and regeneration = reforestation +regeneration [km²/Year]
increase of forest area = IF THEN ELSE (forest area > 100000, reforestation
    −deforestation +regeneration, 0) [km²/Year]
forest area = INTEG (increase of forest area, 3.5e+007) [km²]
deforestation ratio 1990 = DEFORESTATION 1990 /(FOREST AREA 1990
    *AGRICULTURAL AREA 1990) [1/(km²*Year)]

deforestation ratio after 1990 = ANNUAL DEFORESTATION AFTER 1990 /(FOREST AREA 1990 *AGRICULTURAL AREA 1990) [1/(Year*km²)]

deforestation ratio = IF THEN ELSE(Time <= 1990, deforestation ratio 1990, deforestation ratio after 1990) [1/(Year*km²)]

deforestation = forest area *agricultural area *deforestation ratio [km²/Year]

degradation rate = IF THEN ELSE (Time <= 1990, LAND DEGRADATION RATE 1990, LAND DEGRADATION RATE AFTER 1990) [1/Year]

degradation = agricultural area *degradation rate [km²/Year]

increase of agricultural area = deforestation -degradation [km²/Year]

agricultural area = INTEG (increase of agricultural area, 5e+006) [km²]

CO2 release by degradation = degradation *CO2 sequestration difference between fallow and farm land [tCO2/Year]

CO2 sequestration by reforestation = reforestation and regeneration *CO2 sequestration difference between forest and fallow land [tCO2/Year]

CO2 release by deforestation = deforestation *CO2 sequestration difference between farm and forest land [tCO2/Year]

CO2 increase = -(CO2 release by deforestation +CO2 release by degradation +CO2 sequestration by reforestation) /1e+009 [GtCO2/Year]

cumul CO2 from land conversion = INTEG (CO2 increase, 0) [GtCO2]

net CO2 release from land conversion = CO2 increase [tCO2/Year]

CO2 sequestration difference between fallow and farm land = CO2 SEQUESTRATION IN FALLOW LAND -CO2 SEQUESTRATION IN FARMLAND [tCO2/km²]

CO2 sequestration difference between forest and fallow land = CO2 SEQUESTRATION IN FOREST LAND -CO2 SEQUESTRATION IN FALLOW LAND [tCO2/km²]

CO2 sequestration difference between farm and forest land = CO2 SEQUESTRATION IN FARMLAND -CO2 SEQUESTRATION IN FOREST LAND [tCO2/km²]

Simulation time parameters
INITIAL TIME = 1900 [Year]
FINAL TIME = 2100 [Year]
TIME STEP = 0.25 [Year]

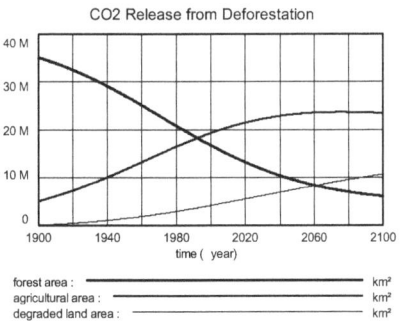

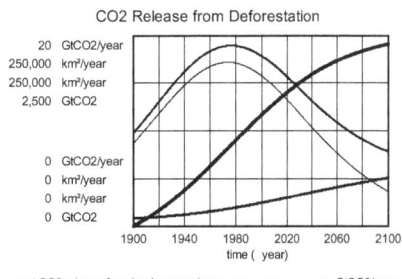

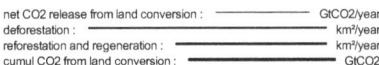

Figure Z304b: Development of forest, agricultural, degraded land area.
Figure Z304c: Deforestation leads to net release of CO₂.

Simulation results

The development of *forest area*, *agricultural area* and *degraded land area* in tropical forest regions, calculated for the period between 1900 and 2100 with the default parameter settings is shown in Figure Z304b. Annual *deforestation* and *reforestation* as well as annual *net CO2 release from land conversion* and *cumulated CO2 from land conversion* are shown in Figure Z304c. Figures Z304d and e show how strongly the key quantities *forest area* and *cumulated CO2 from land conversion* depend on scenario assumptions about the ANNUAL DEFORESTATION AFTER 1990.

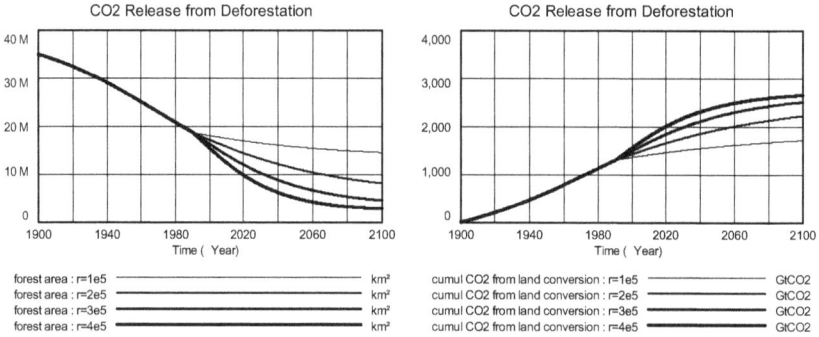

Figure Z304d: Change of forest area for different ANNUAL DEFORESTATION AFTER 1990 (r = 100'000 to 400'000 km^2 per year).
Figure Z304e: Corresponding accumulated CO$_2$ release.

Exercises

1. Use the model to examine the spectrum of possible developments for plausible choices of parameters and scenarios.
2. Compare the results for the rates of CO$_2$ release with the results of the global model Z303 "CO$_2$ dynamics". Which relative contribution does the destruction of tropical forests make to the complete CO$_2$ balance?
3. Using simulations, determine which forest, agricultural and settlement policy measures should be applied to tropical forest areas after the year 2000 to reduce the net emission of CO$_2$ from tropical forest areas to zero as quickly as possible and to possibly use these areas even as CO$_2$ sinks.

References

Deutscher Bundestag 1990:   Schutz der tropischen Wälder – Eine internationale Schwerpunktaufgabe. 2. Bericht der Enquête-Kommission "Vorsorge zum Schutz der Erdatmosphäre". *Zur Sache* 10/1990, Deutscher Bundestag, Bonn.

Z305 Carbon balance of forests

Simulation task

In the process of photosynthesis plants store carbon after absorption of atmospheric CO_2. The amount of CO_2 assimilated by vegetation every year is enormous and far greater than annual CO_2 inputs to the atmosphere by human activities. In an equilibrium state the amount of carbon annually stored in biomass equals the amount annually released to the atmosphere by decomposition of stand litter (cf. models Z302, Z303, and Z304). Even relatively small contributions from human activities can disturb this dynamic equilibrium severely and lead to the climate change.

Likewise, the expansion of forest area (by afforestation etc.) causes additional absorption of CO_2 from the atmosphere, and can contribute to compensation of anthropogenic inputs. International agreements (Kyoto protocol) provide that such contributions can be taken into account as credits in the calculation of national CO_2 balance sheets.

The physiological processes of assimilation of CO_2 by plants and decomposition of organic matter by decomposers are strongly temperature-dependent and correspond to the reaction rate vs. temperature rule of Van't Hoff (Larcher 1980: 103-208, carbon cycle). The net storage of carbon arises as difference of the processes of assimilation and decomposition. It is not immediately obvious under what temperature and other conditions optimal storage arises in which compartments of the ecosystem (biomass, stand litter, humus). Modeling of the system is appropriate for clarification of the total effect of interacting dynamic processes and the resulting CO_2 storage as function primarily of mean annual temperature

Simulation model

The model is documented in the simulation diagrams of Figures Z305a and b and the following model equations. It describes in simplified form (using per hectare values) the dynamics and coupled processes of carbon *assimilation* by photosynthesis, *leaf flushing, stem respiration, wood increment, deadwood loss, litter accumulation, litter decomposition, litter humification, humus mineralization, tree harvest, fuelwood supply* and *timber supply (*with different recycling periods for CO_2). Growth, respiration, and decomposition are strongly temperature-dependent processes. For this reason the stocks of *wood biomass, stand litter* and *humus* as well as the total *CO_2 sequestration in organic matter* in the forest ecosystem are strongly dependent on AVERAGE ANNUAL TEMPERATURE.

The default parameter settings are based on the investigations of Kira 1978 for tropical forests in Southeast Asia. For a far more comprehensive simulation model cf. Bossel 1994, 1996.

Parameters
wood per CO2 = 0.614 [tODM/tCO2]
C per CO2 = 12/44 [tC/tCO2]
AVERAGE ANNUAL TEMPERATURE = 20 [degC]
SPECIFIC RESPIRATION RATE = 0.045 [1/Year]
LITTER DECOMPOSITION RATE = 0.5 [1/Year]
LEAF RESPIRATION RATE NORMAL = 0.42 [1/Year]

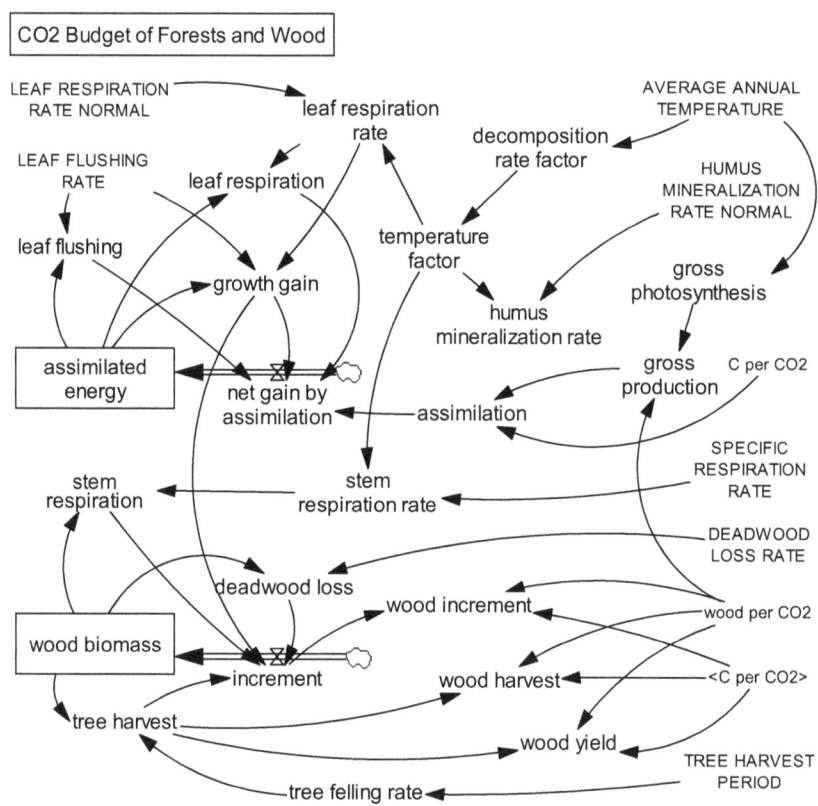

Figure Z305a: Simulation diagram for the carbon balance of forests, part 1.

LEAF FLUSHING RATE = 0.16 [1/Year]
DEADWOOD LOSS RATE = 0.02 [1/Year]
TREE HARVEST PERIOD = 70 [Year]
HUMUS MINERALIZATION RATE NORMAL = 0.01 [1/Year]
HUMIFICATION FRACTION = 0.06 [1]
TIMBER USE DURATION = 50 [Year]
FUELWOOD FRACTION OF TREE HARVEST = 0.2 [1]
WOOD BURN RATE = 1/2 [1/Year]

Growth dynamics
gross photosynthesis = WITH LOOKUP (AVERAGE ANNUAL TEMPERATURE,
    ([(-6, 0) -(40, 100)], (-5, 0), (0, 10), (10, 30), (20, 60), (30, 80), (40, 80)))
    [tODM/(ha*Year)]
gross production = gross photosynthesis /wood per CO2 [tCO2/(ha*Year)]
assimilation = gross production *C per CO2 [tC/(ha*Year)]
net gain by assimilation = assimilation −leaf respiration −leaf flushing -growth gain
    [tC/(ha*Year)]

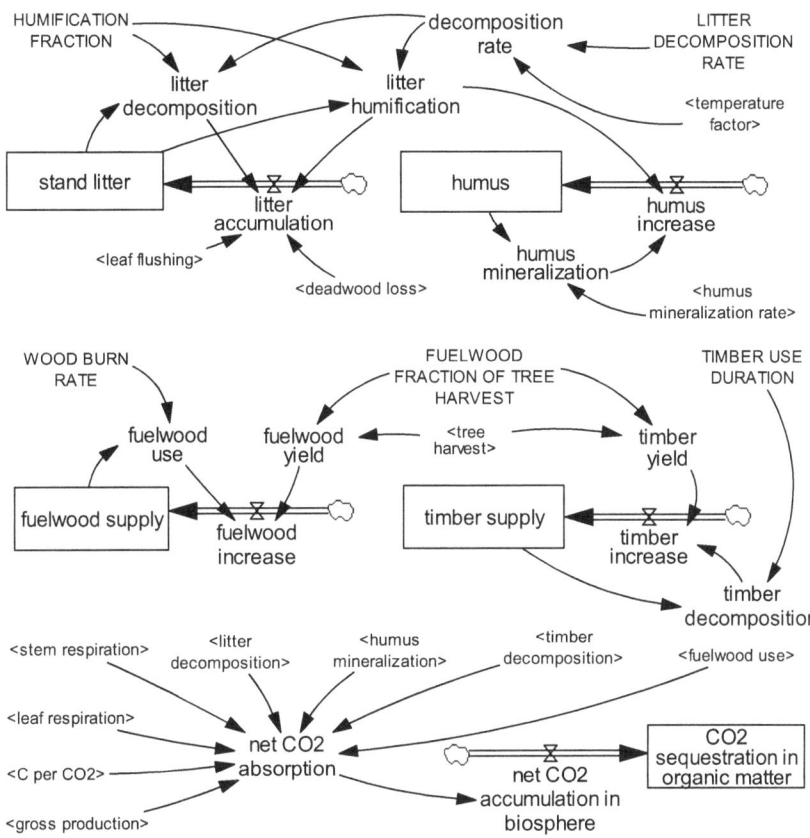

Figure Z305b: Simulation diagram for the carbon balance of forests, part 2.

decomposition rate factor = WITH LOOKUP (AVERAGE ANNUAL TEMPERATURE,
    ([(-6, 0) -(40, 10)], (-5, 0), (0, 0.5), (10, 1), (20, 2), (30, 4), (40, 6))) [1]
temperature factor = decomposition rate factor [1]
humus mineralization rate = HUMUS MINERALIZATION RATE NORMAL
    *temperature factor [1/Year]
leaf respiration rate = LEAF RESPIRATION RATE NORMAL *temperature factor /4
    [1/Year]
leaf respiration = assimilated energy *leaf respiration rate [tC/(ha*Year)]
leaf flushing = assimilated energy *LEAF FLUSHING RATE [tC/(ha*Year)]
growth gain = assimilated energy *(1 –leaf respiration rate -LEAF FLUSHING RATE)
    [tC/(ha*Year)]
assimilated energy = INTEG (net gain by assimilation, 0) [tC/ha]
stem respiration rate = SPECIFIC RESPIRATION RATE *temperature factor /4
    [1/Year]
stem respiration = wood biomass *stem respiration rate [tC/(ha*Year)]
deadwood loss = wood biomass *DEADWOOD LOSS RATE [tC/(ha*Year)]

tree felling rate = 1 /TREE HARVEST PERIOD [1/Year]
tree harvest = wood biomass *tree felling rate [tC/(ha*Year)]
increment = growth gain −stem respiration −deadwood loss -tree harvest
      [tC/(ha*Year)]
wood biomass = INTEG (increment, 0) [tC/ha]
wood increment = increment *wood per CO2 /C per CO2 [tODM/(ha*Year)]
wood yield = tree harvest *wood per CO2 /C per CO2 [tODM/(ha*Year)]
wood harvest = tree harvest *wood per CO2 /C per CO2 [tODM/(ha*Year)]

Decomposition dynamics and CO₂ sequestration
decomposition rate = LITTER DECOMPOSITION RATE *temperature factor [1/Year]
litter decomposition = stand litter *decomposition rate *(1 -HUMIFICATION
      FRACTION) [tC/(ha*Year)]
litter humification = stand litter *decomposition rate *HUMIFICATION FRACTION
      [tC/(ha*Year)]
litter accumulation = leaf flushing +deadwood loss -litter decomposition
      -litter humification [tC/(ha*Year)]
stand litter = INTEG (litter accumulation, 0) [tC/ha]
humus increase = litter humification -humus mineralization [tC/(ha*Year)]
humus mineralization = humus mineralization rate *humus [tC/(ha*Year)]
humus = INTEG (humus increase, 0) [tC/ha]
fuelwood use = WOOD BURN RATE *fuelwood supply [tC/(ha*Year)]
fuelwood yield = FUELWOOD FRACTION OF TREE HARVEST *tree harvest
      [tC/(ha*Year)]
fuelwood increase = fuelwood yield -fuelwood use [tC/(ha*Year)]
fuelwood supply = INTEG (fuelwood increase, 0) [tC/ha]
timber yield = (1 -FUELWOOD FRACTION OF TREE HARVEST) *tree harvest
      [tC/(ha*Year)]
timber decomposition = timber supply /TIMBER USE DURATION [tC/(ha*Year)]
timber increase = timber yield-timber decomposition [tC/(ha*Year)]
timber supply = INTEG (timber increase, 0) [tC/ha]
net CO2 absorption = gross production -(1/C per CO2) *(leaf respiration +stem
      respiration +litter decomposition +humus mineralization +fuelwood use
      +timber decomposition) [tCO2/(ha*Year)]
net CO2 accumulation in biosphere = net CO2 absorption [tCO2/(ha*Year)]
CO2 sequestration in organic matter = INTEG (net CO2 accumulation in biosphere, 0)
      [tCO2/ha]

Simulation time parameters
INITIAL TIME = 0 [Year]
FINAL TIME = 500 [Year]
TIME STEP = 0.25 [Year]

Simulation results

Figure Z305c shows the accumulation of carbon (in tC/ha) as a function of time in
wood biomass, *stand litter*, and *humus* of the forest for the default parameters. The
initial values of these state variables are zero, i.e. the forest starts its growth on soil
lacking any humus initially. *Wood biomass* increases steadily and reaches its equilibrium value after about 75 years. Thereafter, litter fall, deadwood losses, and increment
just balance. *Stand litter* (leaf litter, deadwood) therefore also remains at a constant

level. A flow equilibrium balancing humus formation (*litter humification*) and humus decomposition (*humus mineralization*) develops only after about 200 years; the stock of *humus* still increases during this period.

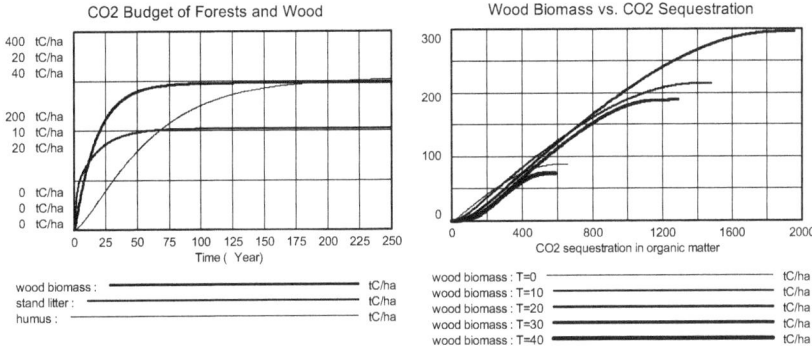

Figure Z305c: Growth of the carbon stock in *wood biomass*, *stand litter*, and *humus*.
Figure Z305d: Mature forests in moderate climate (20 deg C) sequester the greatest amount of carbon.

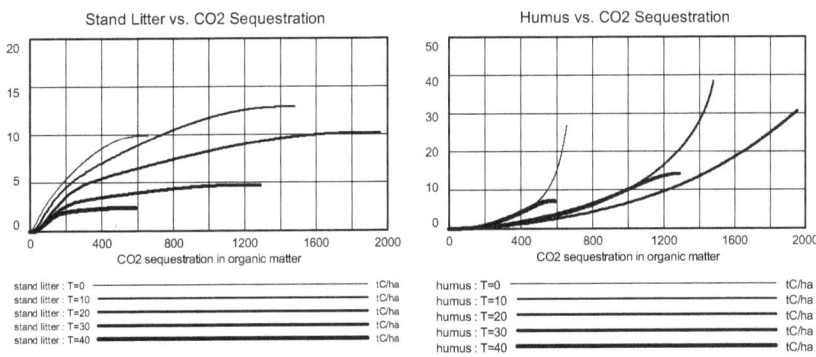

Figure Z305e: *Stand litter* is greater at lower average annual temperatures.
Figure Z305f: The *humus* stock in forest soil is very small for high average annual temperatures.

For conclusions about the possible contribution of forests to the sequestration of CO_2 the temperature influence on the processes of assimilation, respiration, humification, and humus decomposition must be correctly taken into account. State diagrams of the accumulation of carbon (in tC/ha) as *wood biomass, stand litter,* and *humus* as function of total net storage in the forest ecosystem (*CO2 sequestration in organic matter*) are shown in Figures Z305d, e and f. The end points of the curves mark the equilibrium values for these quantities as function of AVERAGE ANNUAL TEMPERATURE (0, 10, 20, 30, 40 degrees centigrade). The following conclusions can be drawn:

1. Optimal carbon sequestration (of almost 2000 tCO_2/ha) occurs at moderate temperatures (here: 20 degrees C). This temperature leads to the most favorable relationship between (temperature dependent) assimilation gains and (also temperature dependent) decomposition losses. Correspondingly, *wood biomass* as well as *stand litter* and *humus* are largest in the respective equilibrium state (mature state of the forest stand).

2. At low temperatures, far more *humus* and *stand litter* accumulates than at higher temperatures, when it decomposes faster.

3. Carbon storage in forests is strongly reduced both in very cold and in very warm climates.

Exercises

1. Use the model to simulate the development of (a) temperate forest (North America, Europe), (b) tropical rain forest, (c) forest in hot desert areas. Compare the development of key quantities for the three cases.

2. What influence does the utilization period of harvested wood have on (a) short-term, (b) long-term sequestration of carbon? Examine the extreme cases: (1) exclusively firewood use, (2) exclusively construction lumber use.

3. Design simplified yet realistic afforestation scenarios for all continents and compute the possible sequestration of atmospheric carbon as function of the mean annual temperatures of the different areas. What contribution could determined reforestation and afforestation make to reduction of the CO_2 increase in the atmosphere? (Compare with the results of model Z303 "CO_2 dynamics of biosphere and atmosphere"). Can reforestation make a contribution in the long run? Do sustainably managed forests contribute to the reduction of the atmospheric CO_2 level?

References

Bossel, H. 1994: *Treedyn3 Forest Simulation Model – Mathematical model, program documentation, and simulation results.* Berichte des Forschungszentrums Waldöko-systeme, Reihe B, Bd. 35, Universität Göttingen.

Bossel, H. 1996: Treedyn3 forest simulation model. *Ecological Modelling* 90, 187-227.

Kira, T. 1978: Community architecture and organic matter dynamics in tropical low-land rainforests in Sourth-East Asia with special reference to Pasoh Forest, West Malaysia. In: P. B. Tomlinson, M. H. Zimmermann: *Tropical Trees as Living Systems.* Proceedings, 4th Cabot Symposium, Harvard Forest, Petersham Mass. Cambridge University Press, Cambridge.

Larcher, W. 1980: *Ökologie der Pflanzen auf physiologischer Grundlage.* UTB Ulmer Stuttgart, 3. Aufl.

Whitmore, T. C. 1985: *Tropical Rain Forests of the Far East.* 2nd ed. Oxford University Press, Oxford UK.

Z306 Motor vehicle traffic and CO_2 emissions

Simulation task

Motor vehicle traffic causes a considerable part of the CO_2 emissions in all countries. In most countries – primarily in fast-developing nations with rapid motorization – fuel consumption and corresponding emissions still grow considerably. The problem is intensified by rapid population growth. This development – with its consequences for the global climate – represents not only an environmental problem but because of rapidly increasing consumption of nonrenewable fossil fuels it also means global conflicts in the long run and a threat to economic and social development. It is therefore important to assess the spectrum of future developments and in particular the consequences of technological developments (such as improvement or deterioration of the fleet efficiency of the vehicle stock) to obtain reliable clues for future-oriented decisions.

Simulation model

The model is documented in the simulation diagrams of Figure Z306a and b and in the following model equations. Population development and development of the gross national product are simulated with simple growth models. Using these results the developments of motor vehicle stock and public transportation are determined. Fuel consumption and CO_2 emissions from private and public transportation are computed taking into account efficiency improvements both in new cars and in the entire car fleet. Public traffic is represented by bus transportation, as is the usual case in many countries.

Population and gross national product (GNP) are calculated assuming logistic growth processes defined by INITIAL POPULATION GROWTH, EXPECTED POPULATION SATURATION LEVEL, INITIAL ECONOMIC GROWTH and MAX GNP PER CAP. The number of cars per cap is determined as exponential saturation process as function of GNP, with parameters determined from international statistical data. The number of cars follows from the size of the population. Service intensity and corresponding service life of cars depend on the relative scarcity of cars (cars per cap). Scrapping rate and purchase rate of new cars are also determined by these factors. Older vehicles have higher fuel consumption than new vehicles. New car efficiency improves in the course of technological development in a logistic process with scenario dependent parameters. The replacement of older vehicles by more efficient new cars leads to fleet efficiency improvement and with that to increase of the car fleet efficiency.

The vehicle transportation performance in person km per day is determined using international empirical data for CAR DRIVING DISTANCE PER DAY and PERSONS PER CAR (Zahavi and Cheslow 1979). Using empirical results (public transportation fraction, Zahavi 1976) the share of public transportation in total transportation of persons (modal split) is determined as a function of motorization (cars per cap). By accounting for average occupancy of vehicles (PERSONS PER CAR, and PERSONS PER BUS) the average performance in terms of car distance per year and bus distance per year can be determined. From these results follow fuel consumption cars and fuel consumption bus, as well as total fuel consumption and total CO_2 emission. The temporal development of these quantities can be examined as function of several parameters that concern efficiency improvements in new vehicles, the share of public transportation in total transportation, and population and economic development.

The structure of the model is generic and is therefore generally valid for countries in different stages of motorization. The default parameter settings apply approximately to a medium-sized industrial country with rapid motorization in the 1960s (Western Germany).

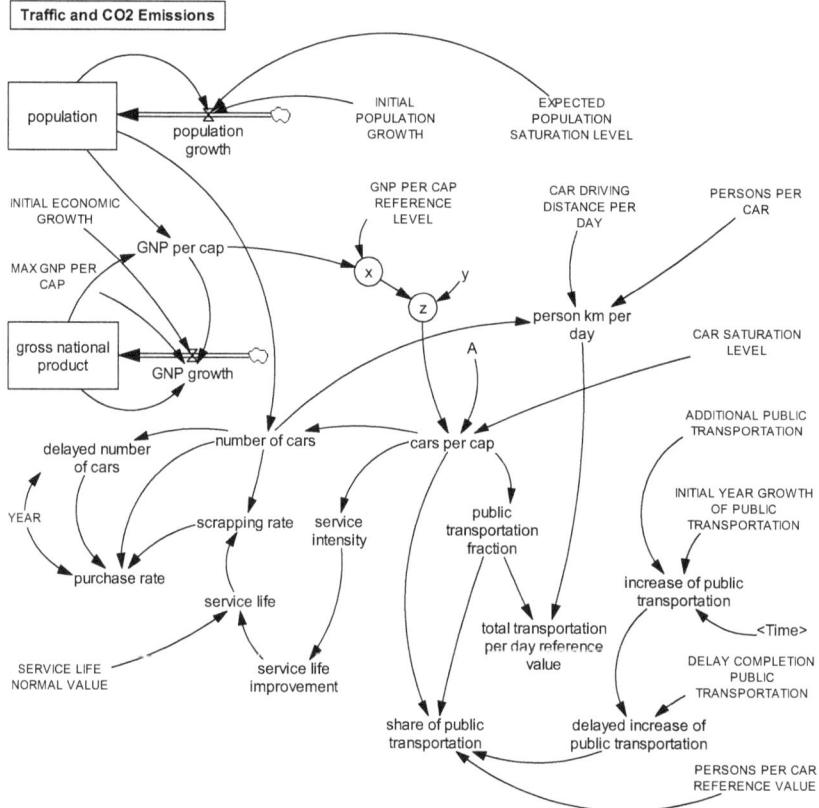

Figure Z306a: Simulation diagram for development of vehicle transportation and CO_2 emissions, part 1.

Parameters
INITIAL POPULATION GROWTH = 0 [1/Year]
EXPECTED POPULATION SATURATION LEVEL = 6e+007 [person]
GNP PER CAP REFERENCE LEVEL = 10000 [$/(person*Year)] *reference level for gross national product per cap, 1985 US*
INITIAL ECONOMIC GROWTH = 0.08 [1/Year]
MAX GNP PER CAP = 50000 [$/(Person*Year)] *GNP/cap saturation level, 1985 $*
CAR DRIVING DISTANCE PER DAY = 30 [km/(car*day)] *average value*

SERVICE LIFE NORMAL VALUE = 7 [Year] *normal car life cycle*
PERSONS PER CAR = 1.2 [person] *average occupancy*
PERSONS PER CAR REFERENCE VALUE = 0.25 [person/car] *reference value for motorization*
CAR SATURATION LEVEL = 1.67 [person/car]
A = 1.2 [1] *initial slope of car saturation curve*
PERSONS PER BUS = 20 [person] *average value*
SPECIFIC FUEL CONSUMPTION BUS = 0.25 [liter/km] *average value*
ADDITIONAL PUBLIC TRANSPORTATION = 0.2 [1]
DELAY COMPLETION PUBLIC TRANSPORTATION = 5 [Year] *delay time*
INITIAL YEAR GROWTH OF PUBLIC TRANSPORTATION = 2000 [Year]
EFFORT FOR EFFICIENCY IMPROVEMENT = 0.5 [1]
MAX EFFICIENCY = 50 [km/liter] *max possible efficiency (min fuel consumption)*
MAX EFFICIENCY RATE = 0.03 [1/Year] *max rate of efficiency improvement*
EFFICIENCY LOSS = 0 [1/Year] *efficiency loss by loss of know-how etc.*
SPECIFIC WEIGHT OF FUEL = 0.6 [kg/liter]
SPECIFIC CO2 EMISSION = 3.3 [1] *CO2 per unit of fuel (kg pro kg)*
DAYS PER YEAR = 365 [day/Year]
YEAR = 1 [Year]

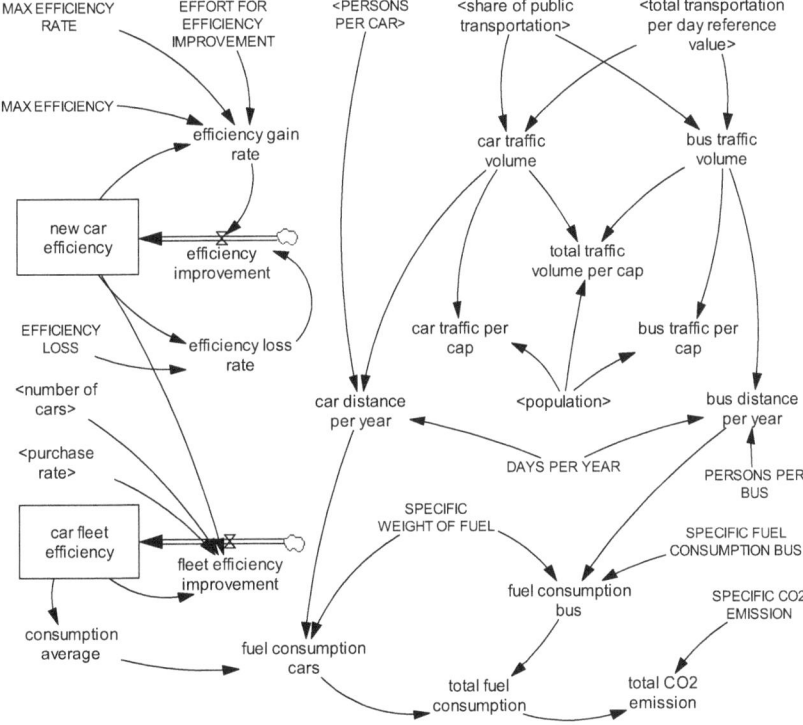

Figure Z306b: Simulation diagram for development of vehicle transportation and CO₂ emissions, part 2.

Dynamics of private cars and public transportation

population growth = INITIAL POPULATION GROWTH *population *(1-population /EXPECTED POPULATION SATURATION LEVEL) [person/Year]

population = INTEG (population growth, 6e+007) [person]

GNP growth = INITIAL ECONOMIC GROWTH *gross national product *(1 -GNP per cap /MAX GNP PER CAP) [$/(Year*Year)]

gross national product = INTEG (GNP growth, 6e+010) [$/Year] *1985 USA*

GNP per cap = gross national product /population [$/(Year*Person)]

x = GNP per cap /GNP PER CAP REFERENCE LEVEL [1]

y = 1.5 [1] *exponent of car saturation function*

z = EXP (y *LN(x)) [1]

cars per cap = (1 -EXP(-A*z)) /CAR SATURATION LEVEL [car/person]

number of cars = cars per cap *population [car]

delayed number of cars = DELAY3 (number of cars, YEAR) [car]

scrapping rate = (1/service life) *number of cars [car/Year]

purchase rate = (number of cars -delayed number of cars) /YEAR +scrapping rate [car/Year]

service intensity = WITH LOOKUP (cars per cap, ([(0, 0) -(2, 5)], (0, 3), (0.05, 2), (0.1, 1.3), (0.3, 0.7), (0.5, 0.5), (1, 0.2), (2, 0.2))) [1] *service intensity as function of car density*

service life improvement = WITH LOOKUP (service intensity, ([(0, 0) -(5, 10)], (0, 0.5), (1, 1), (2, 1.5), (3, 3))) [1] *life cycle improvement for greater service intensity*

service life = service life improvement *SERVICE LIFE NORMAL VALUE [Year]

public transportation fraction = WITH LOOKUP (cars per cap, ([(0, 0) -(2, 1)], (0, 1), (0.03, 0.8), (0.05, 0.67), (0.1, 0.47), (0.2, 0.27), (0.3, 0.15), (0.4, 0.08), (0.5, 0.05), (1, 0.03), (2, 0.025))) [1] *share of public transportation as function of car density*

increase of public transportation = IF THEN ELSE (Time >= INITIAL YEAR GROWTH OF PUBLIC TRANSPORTATION, (1 +ADDITIONAL PUBLIC TRANSPORTA-TION), 1) [1]

delayed increase of public transportation = DELAY3I (increase of public transporta-tion, DELAY COMPLETION PUBLIC TRANSPORTATION, 1) [1]

share of public transportation = public transportation fraction *(1 +(cars per cap /PERSONS PER CAR REFERENCE VALUE) *(delayed increase of public trans-portation -1)) [1]

person km per day = CAR DRIVING DISTANCE PER DAY *number of cars *PERSONS PER CAR [km*person/day]

total transportation per day reference value = person km per day *(1 +(public transpor-tation fraction /(1 -public transportation fraction))) [km*person/day]

Dynamics of fuel consumption

bus traffic volume = total transportation per day reference value *share of public transportation [km*person/day]

bus traffic per cap = bus traffic volume /population [km/day]

bus distance per year = (bus traffic volume /PERSONS PER BUS) *DAYS PER YEAR [km/Year]

fuel consumption bus = bus distance per year *SPECIFIC FUEL CONSUMPTION BUS *SPECIFIC WEIGHT OF FUEL [kg/Year]

car traffic volume = total transportation per day reference value *(1 -share of public transportation) [km*person/day]

car traffic per cap = car traffic volume /population [km/day]

car distance per year = (car traffic volume /PERSONS PER CAR) *DAYS PER YEAR
[km/Year]
fuel consumption cars = car distance per year *consumption average *SPECIFIC
WEIGHT OF FUEL [kg/Year]
total traffic volume per cap = (car traffic volume +bus traffic volume) /population
[km/day]
total fuel consumption = fuel consumption bus +fuel consumption cars [kg/Year]
total CO2 emission = total fuel consumption *SPECIFIC CO2 EMISSION [kg/Year]
efficiency gain rate = ((MAX EFFICIENCY -new car efficiency) /MAX EFFICIENCY)
*MAX EFFICIENCY RATE *EFFORT FOR EFFICIENCY IMPROVEMENT
*new car efficiency [km/(Year*liter)]
efficiency loss rate = EFFICIENCY LOSS *new car efficiency [km/(Year*liter)]
efficiency improvement = efficiency gain rate -efficiency loss rate [km/(Year*liter)]
new car efficiency = INTEG (efficiency improvement, 6.67) [km/liter] *initial value*
100/6.67 = 15 liter/(100 km)
fleet efficiency improvement = (purchase rate/number of cars) *(new car efficiency
-car fleet efficiency) [km/(Year*liter)]
car fleet efficiency = INTEG (fleet efficiency improvement, 5) [km/liter]
consumption average = 1 /car fleet efficiency [liter/km]

Simulation time parameters
INITIAL TIME = 1950 [Year]
FINAL TIME = 2050 [Year]
TIME STEP = 0.2 [Year]

Simulation results

With the default parameter setting rapid motorization beginning in 1950 is simulated.
It leads to a saturation of *number of cars* around the year 2000. The temporal development of the most important quantities is shown in Figures Z306c, d and e.

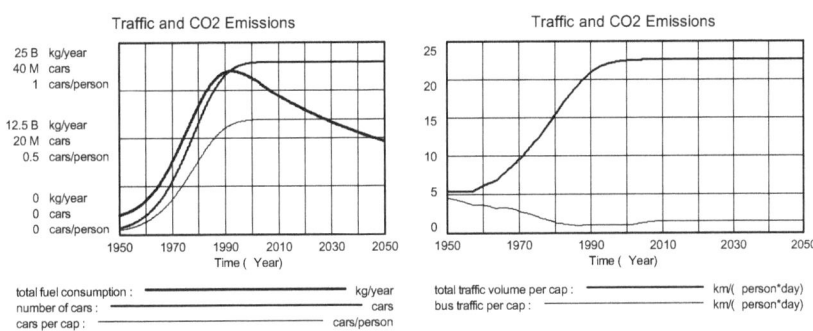

Figure Z306c: Car saturation and fuel consumption for increasing efficiency.
Figure Z306d: The share of public transportation declines strongly.

The number of *cars per cap* increases with increasing *gross national product* until a saturation level is achieved. During this period the transportation shares (modal
split) shift considerably in favor of private car transportation (decrease of *bus traffic*

per cap). At the same time *total traffic per cap* (km car distance per person per day) grows rapidly up to a saturation level.

Since *new car efficiency* improves steadily, *car fleet efficiency* also improves with a corresponding time delay. As a consequence *total fuel consumption* is also reduced in the long term as the *number of cars* saturates.

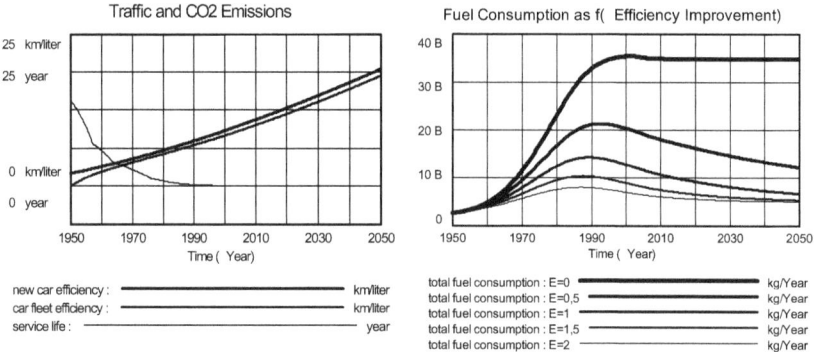

Figure Z306e: Improvement of efficiency for decreasing SERVICE LIFE.
Figure Z306f: EFFORT FOR EFFICIENCY IMPROVEMENT (= *E*) has considerable influence on *total fuel consumption*.

In Figures Z306f, g and h the effects of parameters EFFORT FOR EFFICIENCY IMPROVEMENT (*E* from 0 to 2), CAR SATURATION LEVEL (*P* from 1 to 9 persons per car), and ADDITIONAL PUBLIC TRANSPORTATION (= 2, i.e. 200 percent additional) starting with INITIAL YEAR GROWTH OF PUBLIC TRANSPORTATION (*T* from 1970 to 2010) on *total fuel consumption* are shown. Obviously considerable possibilities exist for reducing fuel consumption and resulting CO_2 emissions.

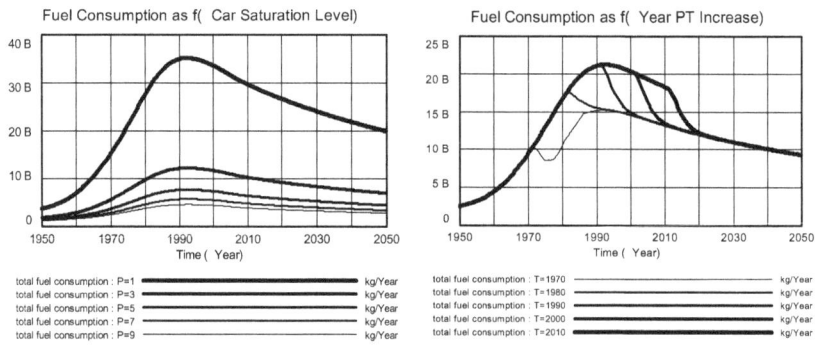

Figure Z306g: Influence of car density saturation on fuel consumption.
Figure Z306h: Influence of a triplication of public transportation on fuel consumption.

Exercises

1. Familiarize yourself with function and dynamics of the model by changing parameters within reasonable limits. Which parameters have the greatest influence on *total fuel consumption* and *total CO_2 emission*?

2. Study the motorization of a developing country with a simultaneously growing population. Use realistic data for a real country, if possible.

3. Investigate, after fitting parameters to the development of a given country, the consequences of speedy introduction of "3-liter cars" (fuel consumption 3 liters per 100 km) on *total fuel consumption* and *total CO2 emission* (using the three parameters determining *efficiency gain rate*).

4. Examine the consequences of a deterioration of *new car efficiency* by a large share of SUVs ("sports utility vehicles") with extremely low fuel efficiency in new car registrations in the USA. Compare this with a speedy introduction of "3-liter cars". By what percentage could total US fuel consumption be reduced by this measure in the long term? Could it make the USA independent of oil and fuel imports?

(*Note:* Sufficiently exact data for these exercises can be found in the internet or in different statistical yearbooks.)

References

Noll, S. A. 1982: *Transportation energy conservation in developing countries*. Resources for the Future, Washington D.C., discussion paper D-73K.

Siddiqi, T. S., Parayno, P., Bossel, H. 1991: Applying system dynamics to climate change issues. *Proceedings, System Dynamics Conference*, Aug. 1991, Bangkok.

Zahavi, Y. 1976: *Travel characteristics in cities of developing and developed nations*. World Bank, Washington D.C., staff working paper no. 230.

Zahavi, Y., Cheslow, M. 1979: Travel demand and estimation of energy consumption by a constrained model. *Transportation Research Record* 764, 79-89.

Z307 Photosynthesis of plants

Simulation task

The photosynthetic production of a stand of plants is dependent on the incident photo-active radiation (PAR) of the sun which in turn depends on the time of day, cloud cover, and seasonally changing position of the sun. During the night no energy assimilation occurs and plants lose energy through respiration. As photosynthesis increases at dawn, the light compensation point is reached where photosynthetic production just covers the respiration losses. With increasing solar radiation the photosynthesis performance of leaves rises, reaching a maximum already at an intermediate radiation value. This light response curve is a characteristic of each plant species.

In the leaf canopy only the topmost leaf layer produces at the maximum rate corresponding to the PAR received. Photoproduction of the lower shaded leaf layers is considerably smaller. For exact statements about the photoproduction of the total leaf canopy a simulation model must therefore account for (1) the diurnal and seasonal dynamics of light radiation; (2) species-specific data (e.g. light response curve); and (3) light attenuation in the different layers of the leaf canopy (mathematical description by Monsi and Saeki 1953).

Simulation model

The model is documented in the simulation diagram of Figure Z307a and in the following model equations. It computes the diurnal dynamics of photosynthetic production of a plant stand as function of geographic latitude, seasonally changing solar declination and day length, diurnal change of solar elevation, and light attenuation in the different layers of the leaf canopy. The model applies to vegetation of all terrestrial ecosystems, i.e. to forests, meadows, fields, bush country etc.

The incident *photoactive radiation* is calculated from GEOGRAPHIC LATITUDE, *solar declination* depending on *time of year,* and *sun elevation* dependent on *time of day,* after accounting for ABSORPTION FACTOR OF ATMOSPHERE, SOLAR CONSTANT and (photoactive) PAR FRACTION IN SUNLIGHT.

The *daily radiation energy* received during the day is the time integral of momentary *photoactive radiation.* The time integral of *light hours* yields the *daylight hours* of a given day.

The net rate of *canopy production* is determined by analytical integration of the leaf production curve as function of LIGHT ATTENUATION and the parameters MAX PHOTOPRODUCTION and SLOPE OF PHOTOSYNTHESIS FUNCTION of leaves, and depending on the number of leaf layers (LEAF AREA INDEX), reduced by AVERAGE LEAF RESPIRATION (Monsi-Saeki equation, cf. France and Thornley 1984, Richter 1985). Integration of *canopy production* over time yields the *daily canopy production* of the canopy layer.

Parameters
CALENDAR DAY = 173 [day]
time of year = (CALENDAR DAY+10) /DAYS PER YEAR [1]
GEOGRAPHIC LATITUDE = 50 [degree]
SOLAR CONSTANT = 1360 [W/m²]

PAR FRACTION IN SUNLIGHT = 0.47 [1]
PAR FACTOR CLOUDED SKY = 1 [1]
ABSORPTION FACTOR OF ATMOSPHERE = 0.15 [1]
LEAF AREA INDEX = 5 [m²/m²]
LIGHT ATTENUATION = 0.7 [1]
MAX PHOTOPRODUCTION = 3 [gCO2/(m²*Hour)]
SLOPE OF PHOTOSYNTHESIS FUNCTION = 0.05 [gCO2/W*Hour]
AVERAGE LEAF RESPIRATION = 0.3 [gCO2/(m²*Hour)]
HOURS PER DAY = 24 [Hour]
DAYS PER YEAR = 365 [day]
PI = 3.14159 [1]
DEGREES PER CIRCLE = 360 [degree]
radian per degree = 2*PI /DEGREES PER CIRCLE [1/degree]
latitude in radian = GEOGRAPHIC LATITUDE *radian per degree [1]

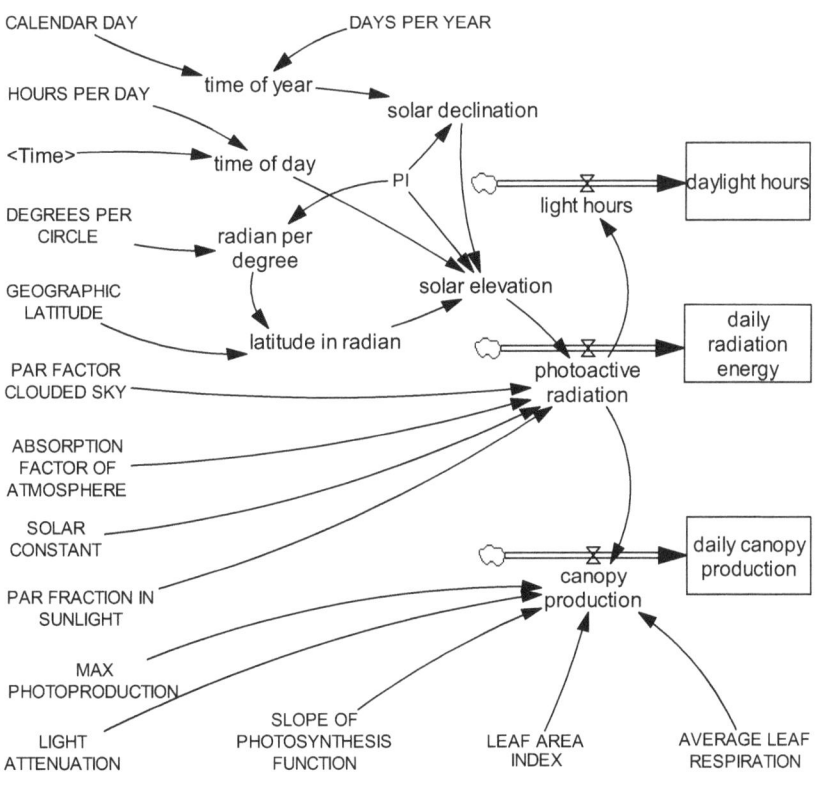

Figure Z307a: Simulation diagram for daily photoproduction of a leaf canopy.

Dynamics
solar declination = -23.4 *(PI/180) *COS (2 *PI *time of year) [1]
time of day = (Time+12) /HOURS PER DAY [1]
solar elevation = SIN (latitude in radian) *SIN(solar declination) +COS(latitude in
    radian) *COS(solar declination) *COS(2 *PI *time of day) [1]
photoactive radiation = IF THEN ELSE(solar elevation > 1/100, PAR FACTOR
    CLOUDED SKY *PAR FRACTION IN SUNLIGHT *SOLAR CONSTANT *solar
    elevation *EXP(-ABSORPTION FACTOR OF ATMOSPHERE /solar elevation),
    0) [W/m²]
daily radiation energy = INTEG (photoactive radiation, 0) [W*Hour/m²]
canopy production = (MAX PHOTOPRODUCTION /LIGHT ATTENUATION)
    *LN((1 +SLOPE OF PHOTOSYNTHESIS FUNCTION /MAX PHOTOPRODUC-
    TION *photoactive radiation) /(1 +(SLOPE OF PHOTOSYNTHESIS FUNCTION
    /MAX PHOTOPRODUCTION) *photoactive radiation *EXP(-LIGHT ATTENUA-
    TION *LEAF AREA INDEX))) -AVERAGE LEAF RESPIRATION *LEAF AREA
    INDEX [gCO2/(m²*Hour)]
daily canopy production = INTEG (canopy production, 0) [gCO2/m²]
light hours = IF THEN ELSE (photoactive radiation > 0.001, 1, 0) [1]
daylight hours = INTEG (light hours, 0) [Hour]

Simulation time parameters
INITIAL TIME = 0 [Hour]
FINAL TIME = 24 [Hour]
TIME STEP = 0.05 [Hour]

Simulation results

Figure Z307b shows results for the 24 hours of day 173 (i.e. June 22nd, summer sol-
stice) and for 50 degrees northern latitude (Frankfurt, Kiew, Vancouver, Winnipeg).
The assumed LEAF AREA INDEX of 5 corresponds to deciduous forest canopy.

    Photoactive radiation follows solar elevation in a roughly sinusoidal fashion
with a maximum at noon. Begin and end of radiation correspond to the times of sun-
rise and sunset. Photoproduction of the leaf layer (*canopy production*) follows the
radiation curve, but shows a fuller profile since maximum leaf photosynthesis occurs
already at intermediate radiation values. Part of daily production is used for respiration
at night; the accumulated production curve (*daily canopy production*) therefore declines during the night. During the summer high values of daily canopy production are
achieved even in polar latitudes as a consequence of the long daylight period there.

    Leaf respiration causes relatively high energy losses if the daylight period is
short. Since lower shaded leaf layers have only small net production, they are usually
dropped, which leads to a maximum LEAF AREA INDEX of about 5 (leaf area index =
square meter of leaf area per square meter of ground area).

    GEOGRAPHIC LATITUDE has a significant effect on production via the dependence
of radiation on solar declination and on season. For total production the shape of the
light response curve of leaves is important: i.e. MAX PHOTOPRODUCTION at the light
saturation point, and the initial SLOPE OF PHOTOSYNTHESIS FUNCTION which charac-
terizes the production increase for increasing radiation (and therefore influences
strongly the production under low radiation conditions at dawn and dusk and in the
lower levels of the leaf canopy).

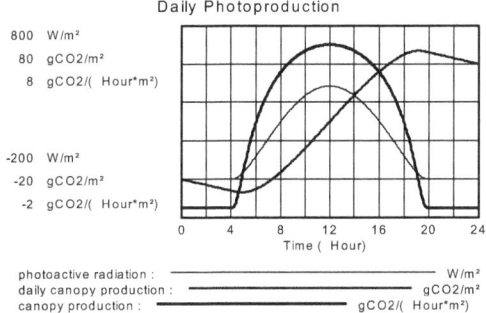

Figure Z307b: Daily photosynthesis of a leaf canopy.

Exercises

1. Determine the production of the plant stand on other days of the year as well as for other geographic latitudes. How is it possible that for equal length of the vegetation period (i.e. 100 days) plant production in temperate and subpolar latitudes can be significantly higher during the summer in these regions than in the tropics? Compute some examples.

2. Determine for the middle of each month the *daily canopy production* (net production) as a function of season, plot its annual course and integrate this time function to find the total annual production (area under the curve). Convert this to annual production of metric tons of carbon per hectare (tC/ha).

3. Explore which net production arises if the LEAF AREA INDEX is changed (in the range from 1 to 10). For which value is net production at a maximum?

4. Examine the effect of AVERAGE LEAF RESPIRATION on net production.

5. Shade-loving plants exhibit a much steeper SLOPE OF PHOTOSYNTHESIS FUNCTION (initial slope of the light response curve) than light-loving plants. Explain the reason and the consequences for the plant. Run simulations for the two plant groups and compare and discuss the result.

Literaturhinweis

France, J., Thornley, J. H. M. 1984: *Mathematical Models in Agriculture*. Butterworths, London (bes. S. 114-137).

Larcher, W. 1980: *Ökologie der Pflanzen auf physiologischer Grundlage*. UTB Ulmer Stuttgart, 3. Aufl.

Richter, O. 1985: *Simulation des Verhaltens ökologischer Systeme – Mathematische Methoden und Modelle*. VCH Weinheim, S. 164-172.

Z308 Forest dynamics

Simulation task

The leaf mass of a forest canopy continues to increase until the canopy has becomes so dense that net production ceases in lower leaf layers where radiation is insufficient due to multiple shading, and new leaves cannot develop (cf. model Z307 "Photosynthesis"). A large portion of the energy assimilated in the canopy is used for respiration to maintain the life processes of the trees. Surpluses are used to produce wood increment and corresponding growth of the trees. The energy demand for respiration increases as (individual) tree biomass increases, and wood increment vanishes when energy gains just compensate energy losses. Obviously the net balance of energy flows (expressed in units of C, CO_2 or organic dry matter ODM) therefore determines the development of a forest stand. Pollutants can impair the energy balance of forests considerably and lead to increment losses and forest dieback by different mechanisms: (1) reduction of photosynthesis, (2) damage of energy assimilating leaves, or (3) by increased photosynthate demand for the replacement of leaf and/or feeder root losses. These interrelated effects can be examined with a simple model describing the dynamic interaction of leaf biomass and wood biomass with their respective energy gains and losses.

Simulation model

The model is documented in the simulation diagram of Figure Z308a and in the following model equations. It has two state quantities: *leaf biomass* and *wood biomass*. All quantities are expressed per hectare of forest area; i.e. the model does not describe development of individual trees.

Logistic growth up to LEAF VOLUME CAPACITY is assumed for *leaf biomass*, with (initial) growth corresponding to MAX LEAF FLUSHING RATE. The limitation of *leaf biomass* (per hectare!) arises from the fact that further increase would reduce radiation on, and photosynthesis of lower leaf layers to a point where the small gains cannot compensate for respiration losses (cf. model Z307 "Photosynthesis"). The production of photosynthates (assimilates) (*canopy production*) is proportional to *leaf biomass* and SPECIFIC CANOPY PRODUCTION; it can be subject to PRODUCTION LOSS BY POLLUTANTS (in percent) which starts in INITIAL YEAR OF ENVIRONMENTAL STRESS. The LEAF PROPORTIONAL ENERGY UTILIZATION accounts for the respiration share of leaves and feeder roots. Further consumption of assimilates (*stem respiration* with STEM PROPORTIONAL RESPIRATION RATE) is proportional to (sap) wood biomass and to leaf flush (*leaf goal*). Losses of *leaf biomass* arise from *leaf loss*; losses of *wood biomass* occur from *deadwood loss* with DEADWOOD LOSS RATE. Any surpluses remaining after accounting for these different losses contribute to *wood increment*.

Parameters
INITIAL WOOD BIOMASS = 1 [t OTS/ha] (*t ODM = metric tons of organic dry matter)*)
INITIAL LEAF BIOMASS = 0.2 [t ODM/ha]
LEAF VOLUME CAPACITY = 10 [t ODM /ha]
MAX LEAF FLUSHING RATE = 0.5 [1/Year]
LEAF LOSS RATE = 0.2 [1/Year]

SPECIFIC CANOPY PRODUCTION = 6 [t ODM /(t ODM *Year)] *canopy production*
     per year relative to leaf mass
LEAF PROPORTIONAL ENERGY UTILIZATION = 0.5 [1]
STEM PROPORTIONAL RESPIRATION RATE = 0.03 [1/Year]
DEADWOOD LOSS RATE = 0.01 [1/Year]
INITIAL YEAR OF ENVIRONMENTAL STRESS = 40 [Year]
PRODUCTION LOSS FROM POLLUTANTS = 0 [1]

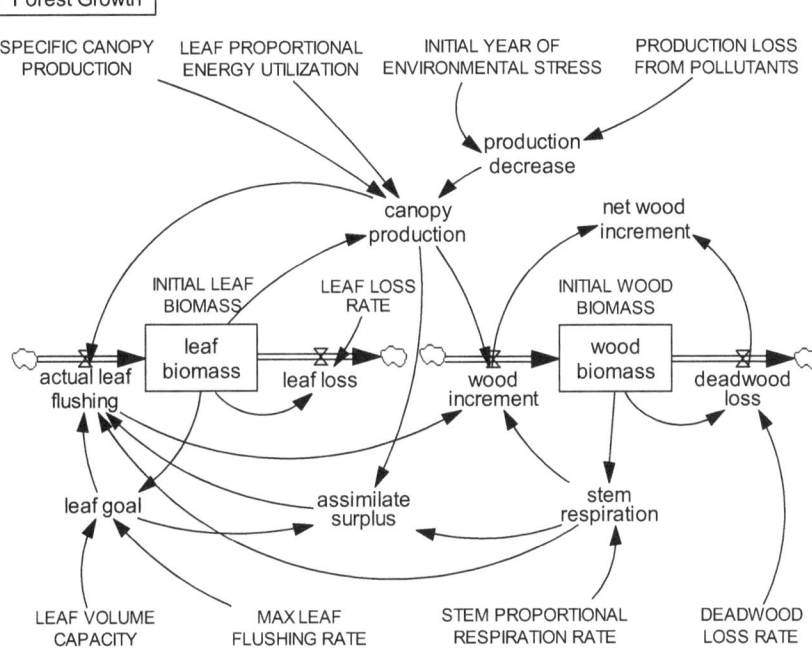

Figure Z308a: Simulation diagram of the forest dynamics model.

Dynamics
production decrease = STEP (PRODUCTION LOSS FROM POLLUTANTS/100,
     INITIAL YEAR OF ENVIRONMENTAL STRESS) [Dmnl]
canopy production = LEAF PROPORTIONAL ENERGY UTILIZATION *leaf biomass
     *SPECIFIC CANOPY PRODUCTION *(1 -production decrease) [t ODM
     /(Year*ha)]
assimilate surplus = canopy production –leaf goal -stem respiration [t ODM /(Year*ha)]
leaf goal = MAX LEAF FLUSHING RATE *leaf biomass *(1 -leaf biomass
     /LEAF VOLUME CAPACITY) [t ODM /(Year*ha)]
actual leaf flushing = IF THEN ELSE (assimilate surplus > 0, leaf goal, IF THEN ELSE
     ((canopy production -stem respiration) > 0, canopy production -stem respiration,
     0)) [t ODM /(Year*ha)]
leaf loss = LEAF LOSS RATE *leaf biomass [t ODM /(Year*ha)]
leaf biomass = INTEG (+actual leaf flushing -leaf loss, INITIAL LEAF BIOMASS)
     [t ODM /ha]

stem respiration = STEM PROPORTIONAL RESPIRATION RATE *wood biomass
     [t OTS/(Year*ha)]
wood increment = canopy production -actual leaf flushing -stem respiration
     [t OTS/(Year*ha)]
deadwood loss = DEADWOOD LOSS RATE *wood biomass [t OTS/(Year*ha)]
wood biomass = INTEG (+wood increment -deadwood loss, INITIAL WOOD
     BIOMASS) [t ODM /ha]
net wood increment = wood increment -deadwood loss [t ODM /(Year*ha)]

Simulation time parameters
INITIAL TIME = 0 [Year]
FINAL TIME = 100 [Year]
TIME STEP = 0.02 [Year]

Simulation results

Figure Z308b shows a time plot of simulation results using the default parameter set, i.e. without pollutant effects. Starting with small initial values for *leaf biomass* and *wood biomass*, the *leaf biomass* quickly grows up to its logistic saturation limit (LEAF VOLUME CAPACITY). As respiration losses proportional to wood biomass (*stem respiration*) are initially small, *net wood increment* reaches an early maximum before decreasing to zero as *stem respiration* and *deadwood loss* increase in proportion to wood biomass and finally all available assimilate is required for maintenance. With the chosen parameters the highest *wood increment* is obtained after about 20 years. In the following years it decreases again because of increasing *stem respiration*. The system therefore approaches an equilibrium state which is defined by the limited production capacity of the leaf canopy.

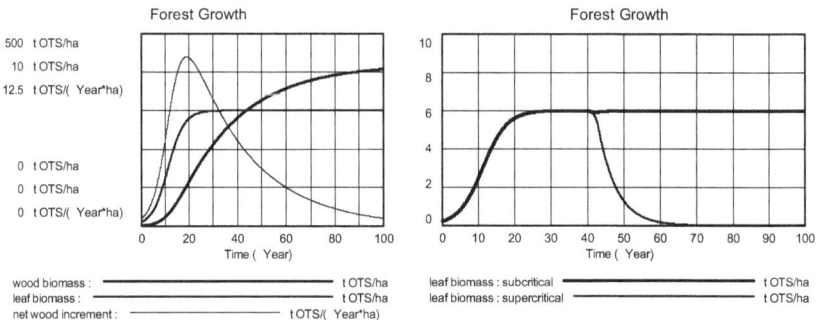

Figure Z308b: Development of a forest stand up to maturity. Wood increment is greatest in the young stand.
Figure Z308c: If a critical pollutant load is exceeded (supercritical stress), the forest suddenly dies. Subcritical damage is difficult to recognize from the external appearance of trees.

The development dynamics is strongly impaired by environmental damage (PRODUCTION LOSS FROM POLLUTANTS) after the INITIAL YEAR OF ENVIRONMENTAL

STRESS. Collapse of the forest can occur if replacement demand for leaves and fine roots as well as respiration demand finally cannot be met any more by energy assimilation. Of critical importance is therefore also SPECIFIC CANOPY PRODUCTION in relation to STEM-PROPORTIONAL RESPIRATION RATE as well as LEAF-PROPORTIONAL ENERGY UTILIZATION as ratio of net assimilation to gross production of leaves.

If a PRODUCTION LOSS FROM POLLUTANTS beginning in INITIAL YEAR OF ENVIRONMENTAL STRESS is introduced, *wood increment* is reduced, but the outer appearance of the forest (*leaf biomass, wood biomass*) hardly changes since a full leaf canopy is still maintained despite decreasing increment. The forest can continue its existence with no increment at all. Only if not enough assimilate can be produced for maintaining sufficient leaf mass the system will rapidly collapse. This dramatic bifurcation of system behavior in response to a minute parameter change is shown in Figure Z308c. For this investigation (with INITIAL YEAR OF ENVIRONMENTAL STRESS = 40) the PRODUCTION LOSS FROM POLLUTANTS was gradually increased. For a value of 46.8 percent the *leaf biomass* still corresponds to that without any pollutant stress – damage could not be recognized from outer appearance. But if the PRODUCTION LOSS FROM POLLUTANTS is only insignificantly increased (to 46.9 percent), the *leaf biomass* disappears in short time and the forest dies. The dynamic process appearing in the model is probably responsible for the process of "forest dieback" occurring in forests stressed by supercritical pollutant load. The more complex model Z309 "Tree dieback" also shows that the assimilate balance (i.e. energy balance) is decisive. It can become negative and lead to collapse either by a production deficit caused by pollution or by increased assimilate demand for renewal of leaves and/or feeder roots due to damage by pollutants.

The global behavior of this simple forest model is examined in the state diagram of Figure Z308d. This diagram was produced by coupling the present model to module Z115 "State diagram" (in Bossel 2007: *System Zoo 1*). The state diagram shows that the "forest" continues to exist and approach an equilibrium point of 6 t$_{ODM}$/ha of *leaf biomass* at about 440 t$_{ODM}$/ha of *wood biomass* if *leaf biomass* is initially large enough in relation to *wood biomass* (left hand upper part of the diagram). If instead the initial *leaf biomass* is relatively too small, the "forest" dies (right hand lower part of the diagram).

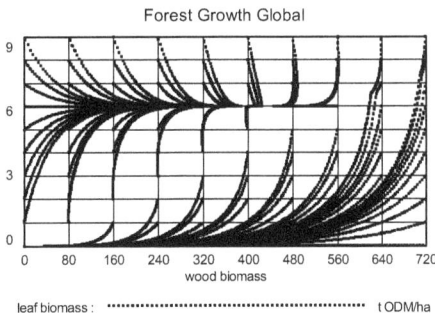

Forest Growth Global

leaf biomass : ··· t ODM/ha

Figure Z308d: The state diagram shows motion of state trajectories toward an equilibrium point in the upper left hand region. Initial values in the lower right hand region lead to collapse.

Exercises

1. Explore the influence in particular of the parameters SPECIFIC CANOPY PRODUC-
TION, LEAF-PROPORTIONAL ENERGY UTILIZATION, STEM-PROPORTIONAL RESPIRATION
RATE.
2. Change the PRODUCTION LOSS FROM POLLUTANTS (percent) and determine up to
which stress values the system can exist without collapse for different INITIAL YEARS
OF ENVIRONMENTAL STRESS. What type of forest can cope better with environmental
stress: younger or older forest?
3. What measures can be taken to protect critically damaged forest stands of different
age from collapse? Investigate this question with different simulation runs by replac-
ing the parameters INITIAL YEAR ENVIRONMENTAL STRESS and PRODUCTION LOSS
FROM POLLUTANTS by time-dependent pollution scenarios defined by table function.
4. Couple module Z115 "State diagram" to the forest dynamics model. Check correct
functioning by reproducing the state diagram of Figure Z308d. (*Note*: Choose the correct state region and computation length (e.g. 20 years) in Z115. Correspondingly,
define FINAL TIME = 100 * 20 = 2000 [years], since 10 * 10 = 100 individual simulation runs are required.) Determine the influence of different pollution scenarios on the
state diagram (INITIAL YEAR OF ENVIRONMENTAL STRESS = 0; different values for
PRODUCTION LOSS FROM POLLUTANTS).

References

Bossel, H. 1986: Dynamics of forest dieback – systems analysis and simulation. *Eco-
logical Modelling* 34 (S. 259-288).
Bossel, H. 1987/1989: *Simulation dynamischer Systeme – Grundwissen, Methoden,
Programme.* Vieweg Braunschweig/ Wiesbaden (S. 245-268).

Z309 Tree dieback

Simulation task

Beginning in the 1980's massive forest dieback episodes were observed in some regions of Central Europe and North America, sometimes leading to total destruction of large areas of forest. Forested highlands, often far from industrial areas were primarily affected, and dieback symptoms occurred both on acidic as well as alkaline soil. While the nature of the epidemic, its causes, its progression, and possible therapies remain uncertain, there is hardly any doubt that air pollution of some kind – and perhaps of different kinds – is the culprit. Pollutants can directly impair the assimilation performance of leaves or else affect tree growth indirectly by soil acidification, dieback of fine roots, nutrient removal, or aluminum or heavy metal poisoning. Despite different causes and physiological consequences in the tree the results are the same: If an overload of stress factors occurs, the tree dies quickly after losing a large part of its leaf canopy and its feeder roots.

The observations have led to the assumption that this "new" type of dieback is a "systemic disease" where the stress of one organ also impairs the function of other organs of a tree until finally the whole tree system collapses. There is much evidence to support this view since the observed symptoms cannot otherwise be linked in a consistent way.

Experience with dynamic systems teaches that system collapse is almost always caused by self-reinforcing feedback mechanisms in connection with nonlinear relationships in the system. The causes of catastrophic collapse are therefore found primarily in structural relationships that come into play if certain parameter constellations are activated. This observation suggests examining the tree system structure and its behavioral dynamics in connection with forest dieback by developing a minimal model of essential relationships in the tree system – perhaps making concessions with respect to quantitative precision – and testing if and under what conditions this model system can produce disastrous collapse. Such a qualitative and explorative model will not produce exact forecasts, but can provide clues concerning possible behavioral dynamics also of the real system.

Simulation model

The simulation diagram of the model for a coniferous tree is shown in Figure Z309a. The model equations are listed in the following. The most important components of the systems are *leaf biomass* and its *assimilate production* by photosynthesis, *root biomass* (fine or feeder roots) and its contribution to nutrient and *water supply*, and finally the distribution of assimilate to the different life processes and organs of the tree. *Remaining assimilate* is used to build up corresponding *wood increment* of the *wood biomass*.

Assimilate production is the product of *leaf biomass* and PHOTOSYNTHESIS EFFICIENCY. The latter may be impaired by pollutants. This is accounted for by appropriate reduction of the standard value "1". The *possible assimilate production* causes a corresponding *feeder root requirement* for providing the necessary water and nutrients. If the available nutrient and *water supply* does not correspond to the *possible assimilate production*, the latter is limited to *actual assimilate production*. The *assimi-*

late requirement for roots is regulated by *feeder root requirement*; the corresponding *assimilate for roots* is used for *root renewal*. *Root loss* is proportional to NORMAL FEEDER ROOT LOSS RATE, but it may increase corresponding to ROOT DAMAGE FACTOR.

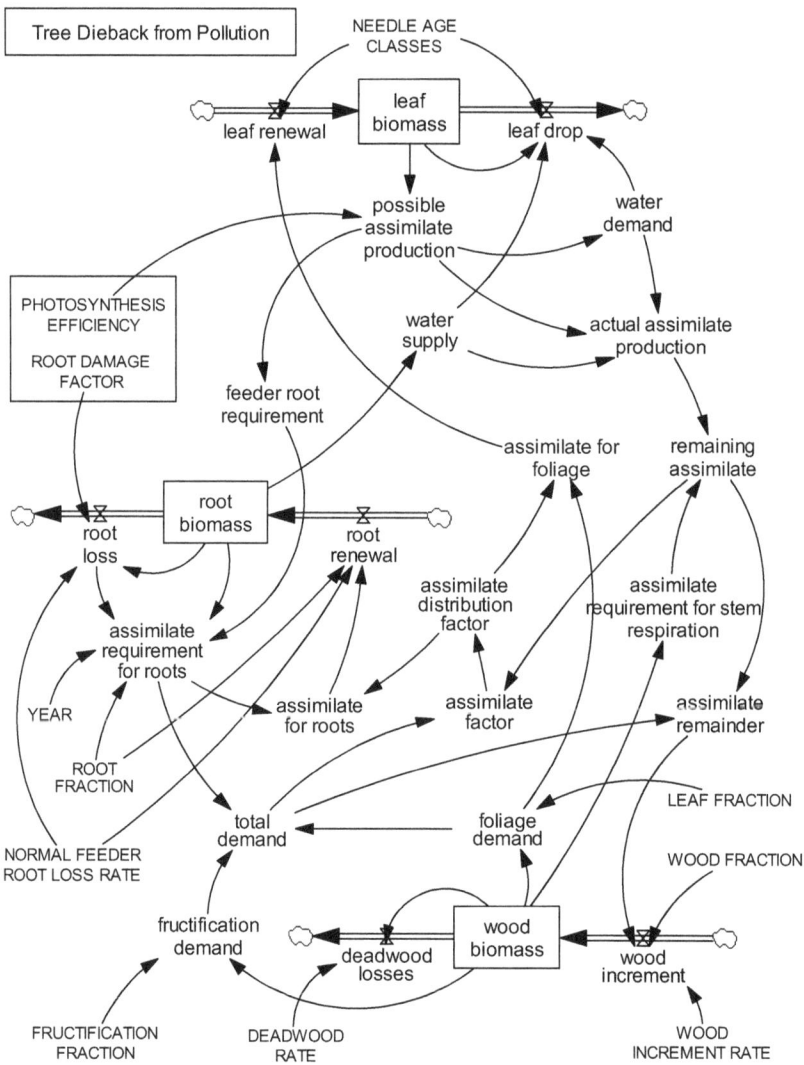

Figure Z309a: Simulation diagram for the model of tree dieback.

Assimilate is distributed according to: *assimilate requirement for stem respiration* (stem, branches, and coarse roots), *assimilate for foliage*, and *assimilate for (fine) roots* and *fructification demand*. The *assimilate remainder* is available for *wood increment*. If sufficient assimilate is not available, then it can be assumed that it is distributed (by *assimilate distribution factor*) in proportion to the different requirements after first deducting current respiration needs. If leaves and/or roots cannot be fully replaced because of assimilate deficiency, collapse occurs by a self-accelerating feedback process.

Variables normalized to "1" were used to quantify the model. For normal development the state variables *leaf biomass*, *root biomass*, and *wood biomass* will therefore be close to unity even over a longer time span. The same applies to *possible assimilate production*. It is reduced to *actual assimilate production* in the ratio of *water supply* to *water demand*. From this amount a maintenance contribution (*assimilate requirement for stem respiration*) corresponding to current *wood biomass* is deducted (weighting 0.3). *Foliage demand* (LEAF FRACTION), *fructification demand* (FRUCTIFICATION FRACTION), as well as *assimilate requirement for roots* (ROOT FRACTION) are deducted from *remaining assimilate*. If this is not sufficient, assimilate distribution is adjusted in proportion while maintaining the relative demand shares. The *assimilate remainder* contributes to *wood increment*.

Details of assimilation, like radiation-dependent photosynthesis, light attenuation in the canopy etc. are not represented in this model. It is therefore not applicable to the simulation of longer-term growth processes of a tree but was designed merely to illustrate the dynamics of collapse. More exact representations of the growth process over a full life cycle use mathematical formulations similar to those of model Z307 "Photosynthesis" (cf. e.g. Bossel 1986, 1989, 1989/1992, 1994, 1996 as well references listed there).

Parameters
NEEDLE AGE CLASSES = 8 [Year]
PHOTOSYNTHESIS EFFICIENCY = 1 [1]
ROOT DAMAGE FACTOR = 1 [1]
NORMAL FEEDER ROOT LOSS RATE = 1 [1/Year]
LEAF FRACTION = 0.15 [1]
ROOT FRACTION = 0.065 [1]
FRUCTIFICATION FRACTION = 0.085 [1]
WOOD FRACTION = 0.075 [1]
WOOD INCREMENT RATE = 1 [1/Year]
DEADWOOD RATE = 0.01 [1/Year]
YEAR = 1 [Year]

Dynamics
possible assimilate production = leaf biomass *PHOTOSYNTHESIS EFFICIENCY [1]
water demand = possible assimilate production [1]
leaf drop = IF THEN ELSE((water demand /water supply) >1.2, leaf biomass
    *(1 /NEEDLE AGE CLASSES) *(water demand /water supply), leaf biomass
    /NEEDLE AGE CLASSES) [1/Year]
water supply = root biomass [1]
actual assimilate production = IF THEN ELSE ((water supply /water demand) < 1,
    possible assimilate production *(water supply /water demand), possible assimi-
    late production) [1]
assimilate requirement for stem respiration = 0.3 *wood biomass [1]

remaining assimilate = actual assimilate production -assimilate requirement for stem
    respiration [1]
assimilate factor = IF THEN ELSE (remaining assimilate >= total demand, 1,
    remaining assimilate /total demand) [1]
assimilate distribution factor = IF THEN ELSE (assimilate factor >= 0, assimilate
    factor, 0) [1]
foliage demand = LEAF FRACTION *wood biomass [1]
assimilate for foliage = foliage demand *assimilate distribution factor [1]
leaf renewal = assimilate for foliage*(1 /NEEDLE AGE CLASSES) *(1 /0.15) [1/Year]
leaf biomass = INTEG (+leaf renewal -leaf drop, 1) [1]
root loss = ROOT DAMAGE FACTOR *NORMAL FEEDER ROOT LOSS RATE
    *root biomass [1/Year]
feeder root requirement = possible assimilate production [1]
assimilate requirement for roots = IF THEN ELSE((root loss *YEAR*(feeder root
    requirement /root biomass) *ROOT FRACTION > 0), root loss *YEAR*(feeder
    root requirement /root biomass) *ROOT FRACTION, 0) [1]
assimilate for roots = assimilate requirement for roots *assimilate distribution factor [1]
root renewal = (assimilate for roots /ROOT FRACTION) *NORMAL FEEDER ROOT
    LOSS RATE [1/Year]
root biomass = INTEG (root renewal -root loss,1) [1]
fructification demand = FRUCTIFICATION FRACTION *wood biomass [1]
total demand = fructification demand +foliage demand +assimilate requirement for
    roots [1]
assimilate remainder = IF THEN ELSE(remaining assimilate >= total demand,
    remaining assimilate -total demand, 0) [1]
deadwood losses = DEADWOOD RATE *wood biomass [1/Year]
wood increment = WOOD FRACTION *assimilate remainder *WOOD INCREMENT
    RATE [1/Year]
wood biomass = INTEG (wood increment -deadwood losses,1) [1]

Simulation parameters
INITIAL TIME = 0 [Year]
FINAL TIME = 10 [Year]
TIME STEP = 0.01 [Year]

Simulation results

The model offers two important possibilities for intervention for simulating pollution
effects. First, a scenario can be entered for the efficiency of assimilation (PHOTOSYN-
THESIS EFFICIENCY). This can be reduced from its normal value "1" to lower values to
represent impairment by pollutants. Second, the loss of fine roots can be simulated by
raising the (normal) rate of *root loss* by a ROOT DAMAGE FACTOR (normal value = 1)
to represent additional root losses by a low pH value (high acidity) or other pollutants
in the soil.

In Figures Z309b, c and d the results from increasing damage to PHOTOSYNTHE-
SIS EFFICIENCY are shown for the three variables *leaf biomass*, *wood biomass* and
wood increment. Normal development (PHOTOSYNTHESIS EFFICIENCY *eff* = 1) shows a
slight increase in *leaf biomass* and *root biomass*, and an almost constant healthy *wood
increment* for more than 10 years. As photosynthesis efficiency is reduced to 0.7, *leaf
biomass* and *root biomass* remain constant while a small *wood increment* is still obtained. Further reduction of *eff* produces similar results: *leaf biomass* and *root biomass*

remain unchanged and *wood increment* reduces to zero. If PHOTOSYNTHESIS EFFI-CIENCY is reduced to 0.55, behavior changes drastically: *Leaf biomass* and *root biomass* reduce to zero quickly, i.e. the tree dies. The system therefore exhibits three distinct behavioral modes: (1) normal growth, (2) stagnation, (3) collapse (Bossel 1986).

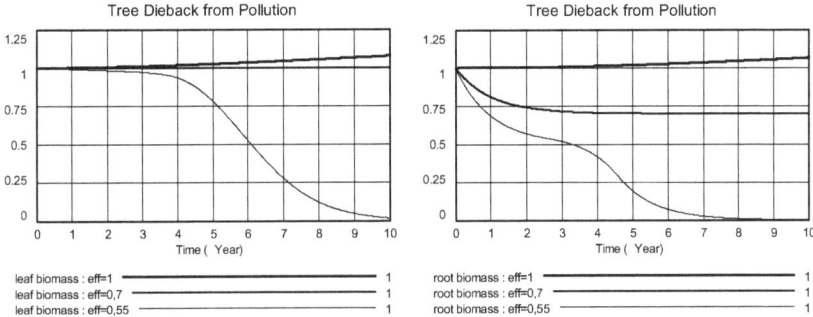

Figure Z309b: Leaf damage resulting from reduction of PHOTOSYNTHESIS EFFICIENCY by pollutants.
Figure Z309c: Root damage following reduction of PHOTOSYNTHESIS EFFICIENCY by pollutants.

Equivalent behavior arises if the ROOT DAMAGE FACTOR (= *root*) is gradually increased (Figure Z309e). At first the resulting greater *root loss* does not seem to affect development. But eventually the system collapses inevitably if a critical damage value is reached (here *root* = 7.18).

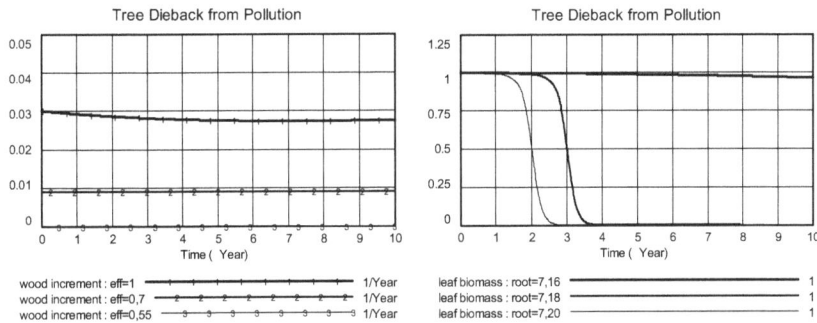

Figure Z309d: Decreasing increment as PHOTOSYNTHESIS EFFICIENCY decreases.
Figure Z309e: Leaf loss and dieback resulting from supercritical root damage.

The symptoms of damage and collapse are the same in both cases (leaf damage or root damage). It does not matter for the tree system whether the primary damage occurs in the leaf system or the root system. Because of the feedbacks in the system, leaf damage also leads to root damage, and *vice versa*. The system finally collapses as a whole. These simulations seem to confirm the assumption of a "systemic disease": Trees can die with identical symptoms either as a result of primary leaf damage or as a result of primary root damage or as a result of a combination of damages in the two organs.

Exercises

1. Experiment with the two damage factors individually and in combination and determine the critical parameter value pairs (PHOTOSYNTHESIS EFFICIENCY, ROOT DAMAGE FACTOR) for which collapse occurs. Plot these value pairs as stability limit in a diagram having the two parameters as axes. Interpret the result.
2. Insert table functions for the damage parameters (PHOTOSYNTHESIS EFFICIENCY, ROOT DAMAGE FACTOR) to allow simulation of time-dependent damage scenarios.
3. In agreement with empirical observations the simulation results indicate that collapse has to be expected if little or no wood increment is produced. What does this mean for possible early detection of endangered forest stands? What possibilities for prevention of a beginning collapse exist, if at all? Find out by simulations how quickly and strongly ecological damage must be reduced to avoid imminent collapse.
4. Obtain data for a specific tree species and replace the normalized quantities in the model by absolute values and species-specific assimilate partitioning. What results do you obtain? Can you relate the results to empirical observations?
5. Expand and supplement the model by using real (dimensional) quantities and a more exact representation of assimilation in the leaf canopy (cf. model Z307 "Photosynthesis" or Bossel/Schäfer 1989) to allow simulation of tree development over the full life cycle. Compare the simulation results with empirical data.

References

Bossel, H. 1985: *Umweltdynamik – 30 Programme für kybernetische Umwelterfahrungen*. Te-wi, München (S. 173-195).

Bossel, H., Metzler, W., Schäfer, H. 1985: *Dynamik des Waldsterbens – Mathematisches Modell und Computersimulation*. Springer Verlag, Berlin, Heidelberg, New York, Tokyo 1985.

Bossel, H. 1986: Dynamics of forest dieback – Systems analysis and simulation. *Ecological Modelling*, 34 (S. 259-288).

Bossel, H., Schäfer, H. 1989: Generic simulation model of forest growth, carbon and nitrogen dynamics, and application to tropical acacia and European spruce. *Ecological Modelling* 48, S. 221-265.

Bossel, H. 1989/1992: *Simulation dynamischer Systeme – Grundwissen, Methoden, Programme*. Vieweg Braunschweig/ Wiesbaden (S. 245-268).

Bossel, H. 1994: *Treedyn3 Forest Simulation Model – Mathematical model, program documentation, and simulation results*. Berichte des Forschungszentrums Waldökosysteme, Reihe B, Bd. 35, Universität Göttingen.

Bossel, H. 1996: Treedyn3 forest simulation model. *Ecol. Modelling* 90, 187-227.

Z310 Soil water dynamics

The role of simulations in agriculture

Agriculture is characterized by dynamic processes of growth and decay. Some have a typical time span of days (e.g. pest invasions), some of weeks (e.g. plant growth up to harvest) or of months (e.g. livestock breeding), some of years (soil erosion, growth of forests, rural structural change). In these dynamic processes many mutually connected ecological, technological, economic, and management variables and decisions often influence each other in complex feedback patterns. Often it becomes difficult to reliably predict the consequences of a decision, and the system manager (the farmer) must rely on experience and intuition. This often leads to situations where high risks must be taken (as in decisions for pesticide applications) or unnecessary waste of resources occurs (fuels, fertilizers, chemicals, water, labor, soil).

A better understanding of dynamic systems, their processes and characteristics can lead to better understanding of the critical components of a system and of more effective and less resource intensive options of management and control. The effort to better understand system relationships, to formalize them in mathematical terms, and to capture them in a dynamic simulation model whose behavior can be investigated under a variety of different conditions can effectively lead to better system understanding and to better decisions in many cases.

The objective for developing the three models introduced here for field crop cultivation (Z310, Z311, Z312) was not to provide programs for the management of crop production. The aim was rather to include and connect in a model as small as possible those elements of the dynamic system which can provide a valid description of essential system structure and resulting system dynamics. During development of the model the emphasis therefore lay on identifying the most important elements and their interconnections and not on precision or detail of the representation. It cannot be expected therefore that the simulation results describe specific local conditions exactly, but one can expect a fairly reliable description of time response even for a wide variety of parameter values.

In this qualified sense the models introduced here must be understood as didactic models, i.e. models which facilitate and improve understanding of relatively complex systems and their behavior under a variety of conditions. If models of this type can provide a contribution to better decisions, this contribution would come from resulting better system understanding and not from direct application to a concrete problem. Models of this type can convey very much more information about a system than is possible by a verbal description. In particular, a broad spectrum of conditions and decision options can be examined with very little effort with such models. The results will provide the model user with a much better "sense" of the system and its behavior. This understanding will enable the model user to make better informed decisions also in the real system under his management.

Relatively aggregate models like the present one can also be understood as initial and simple prototypes for much more complex, data-intensive, and accurate models for crop production management. The fundamental simulation approach will remain identical for many applications. (We do not deal here with a type of model which plays a certain role in agricultural management and which is based on a completely different approach – i.e. optimization models used to optimize amount and composition of feed in feedlots or poultry farms, for example.) The models introduced here can be developed into more detailed models without decisive changes in model approach or model design.

Models concerning agricultural issues are conceivable at different levels: A first level is that of national or regional agricultural production, a second that of the individual farm, a third that of plant production in a particular field with given soil parameters, and a fourth the level of the individual plant or the individual animal. The following three models deal with crop cultivation at the field level.

The first model Z310 "Soil water dynamics" is a representation of the dynamic changes of soil moisture in a field. In this model soil moisture is determined as a function of precipitation, percolation, soil conditions, and growth and transpiration of plants.

The second model Z311 "Nutrient dynamics" also works at field level and has been developed for coupling it to the model of soil water dynamics. It concentrates on the dynamics of soil nitrogen as a function of nutrient uptake, fertilization, decomposition of organic material, leaching etc. In this model the characteristic parameters of different field crops can be introduced, allowing simulation of the nitrogen balance under different conditions and for prescribed crop rotation schemes.

The third model Z312 "Field crop production" was constructed by linking models Z310 "Soil water dynamics" and Z311 "Nutrient dynamics". It allows the simultaneous simulation of the dynamics of soil water supply, nutrient availability, and plant growth. Among other things, the model can be used for initial crop rotation studies.

Simulation task for soil water dynamics

Growth of a field crop and crop yield depend fundamentally on soil water availability. It is a function of the dynamic processes of precipitation, evaporation, irrigation, percolation, drainage, transpiration (of the crop) and – most importantly – of soil parameters. In the course of a year, soil moisture constantly changes as a function of changing conditions. Soil moisture has a decisive effect on the dynamics of plant growth, which in turn determines transpiration and hence soil water supply. Formalization of the model must account for a large number of soil and plant parameters such as: soil type (clay content), depth of plow layer (topsoil), organic matter content, root depth, compaction by agricultural machinery, ground cover (plants or plastic foils), mineral fertilizer inputs, crop type, times of sowing, planting and harvesting etc. It should be possible to modify and change total annual precipitation and precipitation pattern to allow examining their influence on soil water availability and crop yield.

The model introduced here deals essentially with soil water availability. Its purpose is representation of the major relationships determining water availability in farm fields and the computation of their dynamic consequences as a function of a multitude of soil, weather and plant parameters to be entered interactively.

In a first step the model calculates the water holding capacity of the soil from given soil parameters (clay fraction, organic matter content, depth of plow layer, root depth, soil compaction). Plant growth, transpiration, evaporation, and resulting soil water availability in the course of the vegetation period are determined from parameters specified by the user for precipitation (with random distribution), initial soil water content, ground water level (if within capillary rise), irrigation scenario, planting and

harvesting dates, and maximum biomass. If the soil water supply is insufficient, the
plants wilt and die.

There is only a summary description of the nutrient supply. The model can therefore only serve as a submodel of a more complete representation of field crop growth.
It was designed for linkage to model Z311 "Nutrient dynamics" discussed in the following contribution. It is instructive, however, to use the isolated model to examine
the influence of different conditions on soil water availability over the span of a year.
The interactive use of the model allows selecting and changing all essential parameters individually.

Simulation model

The model shall be explained by reference to the simulation diagram (Figures Z310a
and b). The corresponding model equations are also documented completely in the
following.

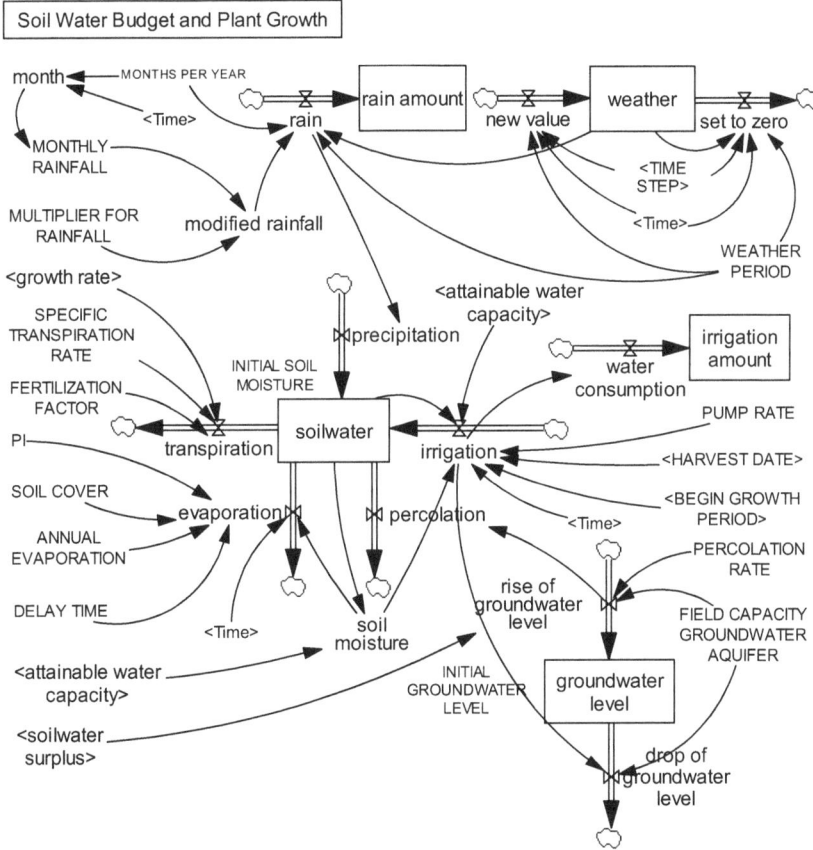

Figure Z310a: Simulation diagram of soil water dynamics, part 1.

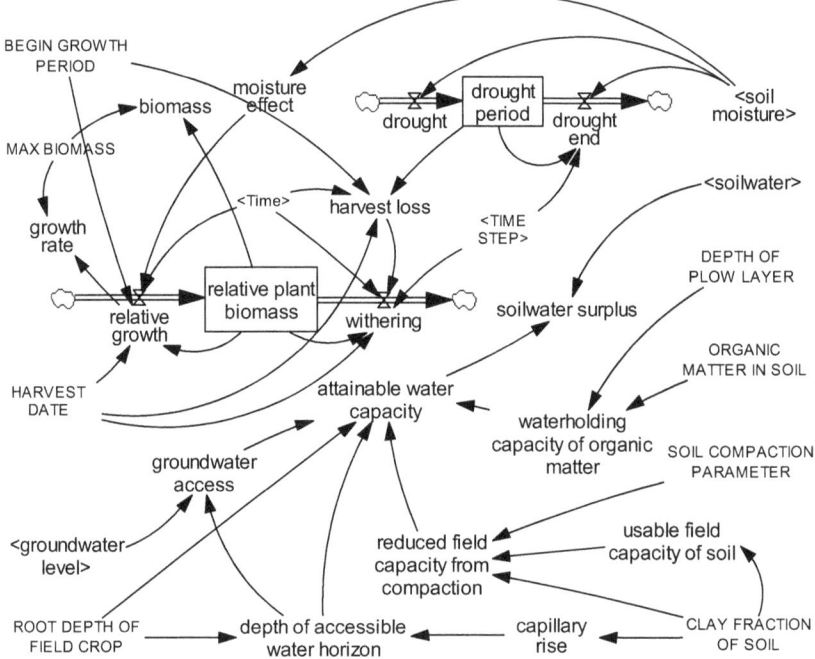

Figure Z310b: Simulation diagram of soil water dynamics, part 2.

The amount of *soil water* increases by *precipitation* and *irrigation*, while *evaporation*, *transpiration* and *percolation* reduce the pool of *soil water*. *Evaporation* and amount of *irrigation* (if irrigation is used at all with PUMP RATE) depend on *soil moisture*. The *irrigation* decision depends also on the vegetation period while *evaporation* is a function of seasonally changing solar radiation and SOIL COVER (bare ground, plant cover, foil).

The amount of plant-available water depends on *soil moisture*. The *relative growth* of *relative plant biomass* changes in response to *moisture effect*. If *soil moisture* decreases below a certain value (wilting point), then plant growth is severely disrupted. The plant dies if the *drought period* continues for several weeks.

Growth starts at the time of sowing or planting (BEGIN GROWTH PERIOD) and ends at HARVEST DATE. The rate of *relative growth* is strongly dependent on already available *relative plant biomass*. Initial rapid growth reduces to zero gradually toward the end of the growth period, following a logistic saturation function. *Relative growth* is converted to (absolute) *growth rate* by multiplying by the MAX BIOMASS to be expected for the assumed nutrient supply (the plant-specific effects of the nutrient supply are represented in model Z311 "Nutrient dynamics"). *Transpiration* depends on the *growth rate* of the crop and partly also on nutrient availability (FERTILIZATION FACTOR). A *harvest loss* arises for inadequate *soil moisture* during a lengthy drought period unless the drought occurs shortly before HARVEST DATE.

The behavior of the system is strongly influenced by soil parameters. The DEPTH OF PLOW LAYER together with the content of ORGANIC MATTER IN SOIL determines the *water holding capacity of organic matter* in the soil. The water holding capacity of the soil itself (expressed by *usable field capacity of soil*) is primarily a function of soil texture. It is modified by SOIL COMPACTION PARAMETER. A good measure of soil texture is the relative amount of clay particles (smaller than 0.01 mm), i.e. the CLAY FRACTION OF SOIL. It also determines the *capillary rise* of the water. The ROOT DEPTH OF FIELD CROP together with *capillary rise* determines the *depth of accessible (ground) water horizon* from which plants can take up water. Together, these different factors determine *attainable water capacity* of the soil. From this and the absolute amount of *soil water* the *soil moisture* can be determined.

If the amount of *soil water* exceeds the water holding capacity (*soil water surplus*), part of the soil water will seep away (*percolation*) and increase the *ground water level*. On the other hand, the *ground water level* decreases as a function of the FIELD CAPACITY OF GROUND WATER AQUIFER and the amount of ground water pumped out for *irrigation*.

For a simulation run the default settings for precipitation data (corresponding to annual *rain amount* of about 800 mm/year) can be modified using the MULTIPLIER FOR RAINFALL, or the table function for MONTHLY RAINFALL can be used to specify a particular rainfall scenario. Rainfall episodes can be determined by a random function, where the average length of the typical WEATHER PERIOD must be specified.

Data and table functions used in the model are based on data from the technical literature (cf. references).

Soil parameters
DEPTH OF PLOW LAYER = 0.3 [m]
ORGANIC MATTER IN SOIL = 0.08 [1] *volume fraction*
CLAY FRACTION OF SOIL = 0.3 [1] *volume fraction*
FIELD CAPACITY GROUNDWATER AQUIFER = 0.15 [1] *m water column per m soil*
INITIAL SOIL MOISTURE = 0.5 [1]
INITIAL GROUNDWATER LEVEL = -10 [m]
PERCOLATION RATE = 50 [1/Year]
SOIL COVER = 1 [1] *bare soil = 0, plant cover = 1, plastic foil = 2*
SOIL COMPACTION PARAMETER = 0 [1] *soil compaction = 1, no compaction = 0*
usable field capacity of soil = WITH LOOKUP (CLAY FRACTION OF SOIL, ([(0, 0)
    -(1, 1)], (0, 0.01), (0.1, 0.1), (0.2, 0.16), (0.3, 0.2), (0.4, 0.21), (0.5, 0.15), (0.6,
    0.13), (1, 0.05))) [1]
capillary rise = WITH LOOKUP (CLAY FRACTION OF SOIL, ([(0, 0) -(1, 10)], (0, 0.1),
    (1, 3))) [m]

Crop parameters
MAX BIOMASS = 10000 [kg/ha] *biomasse in kg ODM (organic dry matter)*
SPECIFIC TRANSPIRATION RATE = 0.4 [m/(kg/ha)]
ROOT DEPTH OF FIELD CROP = 1 [m]
BEGIN GROWTH PERIOD = 0.3 [Year] *week 13 = 13/52 = 0.25*
HARVEST DATE = 0.6 [Year] *week 26 = 26/52 = 0.5*
FERTILIZATION FACTOR = 1 [1] *optimal fertilization = 1, no fertilization = 0,
    intermediate values OK*
PUMP RATE = 150 [1/Year] *no irrigation = 0*

Weather parameters
MONTHLY RAINFALL = WITH LOOKUP (month, ([(0, 0) -(12, 0.1)], (0, 0.068),
     (1, 0.068), (2, 0.068), (3, 0.047), (4, 0.061), (5, 0.063), (6, 0.075), (7, 0.086),
     (8, 0.09), (9, 0.066), (10, 0.067), (11, 0.068), (12, 0.068))) [m/Month]
     Note: value for 0 = value for 12 (end of year); annual sum = 1 to 12
MULTIPLIER FOR RAINFALL = 1 [1]
ANNUAL EVAPORATION = 0.444 [m/Year] *m WS*
WEATHER PERIOD = 3 [Day] *no random calculation if = 0*

Constants
MONTHS PER YEAR = 12*1 [Month/Year]
month = MONTHS PER YEAR *Time [Month]
PI = 3.14159*1 [1/Year]
DELAY TIME = 0.25/1 [Year]

Precipitation dynamics
new value = IF THEN ELSE (WEATHER PERIOD <= 0, 0, IF THEN ELSE
     (ABS (Time *(365 /WEATHER PERIOD) –INTEGER (Time *(365 /WEATHER
     PERIOD))) < TIME STEP *(365 /WEATHER PERIOD), INTEGER (RANDOM
     UNIFORM (0.5, 1.5, 0)) /TIME STEP, 0)) [1/Year]
set to zero = IF THEN ELSE (WEATHER PERIOD <= 0, 0, IF THEN ELSE
     (ABS ((Time +TIME STEP /2) *(365 /WEATHER PERIOD) –INTEGER ((Time
     +TIME STEP /2) *(365 /WEATHER PERIOD))) < TIME STEP *(365 /WEATHER
     PERIOD), weather /TIME STEP, 0)) [1/Year]
weather = INTEG (+new value -set to zero, 0) [1]
rain = IF THEN ELSE(WEATHER PERIOD <= 0, MONTHS PER YEAR *modified
     rainfall, MONTHS PER YEAR *modified rainfall *weather *2) [m/Year]
rain amount = INTEG (rain, 0) [m]
modified rainfall = MONTHLY RAINFALL *MULTIPLIER FOR RAINFALL [m/Month]
precipitation = rain [m/Year]

Soilwater dynamics
waterholding capacity of organic matter = DEPTH OF PLOW LAYER *ORGANIC
     MATTER IN SOIL *5 [mm WS]
reduced field capacity from compaction = IF THEN ELSE (SOIL COMPACTION
     PARAMETER > 0 :AND: CLAY FRACTION OF SOIL < 0.15, 0.94*usable field
     capacity of soil, IF THEN ELSE (SOIL COMPACTION PARAMETER > 0 :AND:
     CLAY FRACTION OF SOIL >= 0.15, 0.88 *usable field capacity of soil, usable
     field capacity of soil)) [1]
transpiration = (1 /FERTILIZATION FACTOR) *SPECIFIC TRANSPIRATION RATE
     *growth rate /10000 [m/Year]
evaporation = soil moisture *ANNUAL EVAPORATION *(1 +0.95*SIN (2 *PI *(Time
     -DELAY TIME))) *(1 -SOIL COVER /2) [m/Year]
percolation = rise of groundwater level [m/Year]
soilwater = INTEG (irrigation +precipitation -transpiration -evaporation -percolation,
     INITIAL SOIL MOISTURE *attainable water capacity) [mm WS]
soil moisture = soilwater /attainable water capacity [1]
soilwater surplus = soilwater -attainable water capacity [m]
irrigation = IF THEN ELSE (soil moisture < 0.5 :AND: Time -INTEGER(Time) > BEGIN
     GROWTH PERIOD :AND: Time -INTEGER(Time) < HARVEST DATE, (0.5
     *attainable water capacity -soilwater) *PUMP RATE, 0) [m/Year]
water consumption = irrigation [m/Year]

irrigation amount = INTEG (water consumption, 0) [m]
drop of groundwater level = irrigation /FIELD CAPACITY GROUNDWATER AQUIFER
    [m/Year]
rise of groundwater level = IF THEN ELSE (soilwater surplus > 0, (soilwater surplus
    /FIELD CAPACITY GROUNDWATER AQUIFER) *PERCOLATION RATE, 0)
    [m/Year]
groundwater level = INTEG (+rise of groundwater level -drop of groundwater level,
    INITIAL GROUNDWATER LEVEL) [m]

Crop dynamics
depth of accessible water horizon = ROOT DEPTH OF FIELD CROP +capillary rise
    [m]
attainable water capacity = IF THEN ELSE (groundwater access > 0, depth of
    accessible water horizon *reduced field capacity from compaction, waterholding
    capacity of organic matter +ROOT DEPTH OF FIELD CROP *reduced field
    capacity from compaction) [m]
groundwater access = IF THEN ELSE (-depth of accessible water horizon < ground-
    water level, 1, 0) [1]
moisture effect = WITH LOOKUP (soil moisture, ([(0, 0) -(2, 2)], (0, 0), (0.1, 0.2),
    (0.3, 0.5), (0.5, 1), (1, 1), (2, 1))) [1]
relative growth = IF THEN ELSE ((Time –INTEGER (Time) < BEGIN GROWTH
    PERIOD) :OR: (Time –INTEGER (Time) > HARVEST DATE), 0, (25*(20/52)
    /(HARVEST DATE -BEGIN GROWTH PERIOD)) *relative plant biomass
    *(1 -relative plant biomass) *moisture effect) [1/Year]
growth rate = relative growth *MAX BIOMASS [kg/(ha*Year)]
drought = IF THEN ELSE (soil moisture > 0.2, 0, 1) [1/Year]
drought period = INTEG (drought -drought end, 0) [1]
drought end = IF THEN ELSE (soil moisture > 0.2, drought period /TIME STEP, 0)
    [1/Year]
harvest loss = IF THEN ELSE (drought period > 20/365 :AND: (Time -INTEGER(Time)
) > BEGIN GROWTH PERIOD :AND: Time -INTEGER(Time) < BEGIN
    GROWTH PERIOD +(HARVEST DATE -BEGIN GROWTH PERIOD), 1, 0) [1]
withering = IF THEN ELSE ((Time -INTEGER(Time) > HARVEST DATE) :OR:
    (harvest loss = 1), relative plant biomass /TIME STEP, 0) [1/Year]
relative plant biomass = INTEG (relative growth -withering, 0.01) [1]
biomass = relative plant biomass *MAX BIOMASS [kg/ha]

Simulation time parameters
INITIAL TIME = 0 [Year]
FINAL TIME = 1 [Year]
TIME STEP = 0.01 [Year]

Simulation results

The time span of individual simulation runs is one year. The large number of soil, plant and weather parameters allows simulation of very different conditions. Influences of soil parameters, precipitation, and irrigation on soil water availability and plant development are exemplarily examined in the following.

Figure Z310c shows results for the default parameter setting. They apply to relatively deep, humus rich topsoil on uncompacted clay subsoil. The water holding capacity is relatively high therefore, as is the capillary rise (because of small size of clay particles). Since the ground is covered completely by plants, evaporation on the soil surface is reduced correspondingly. Crop parameters correspond approximately to those of a grain crop. Precipitation data and weather parameters approximately represent conditions in the western part of Central Europe. The simulation results show that under these conditions the field crop develops almost to its MAX BIOMASS even without irrigation. The *ground water level* increases slightly by *percolation* of soil water.

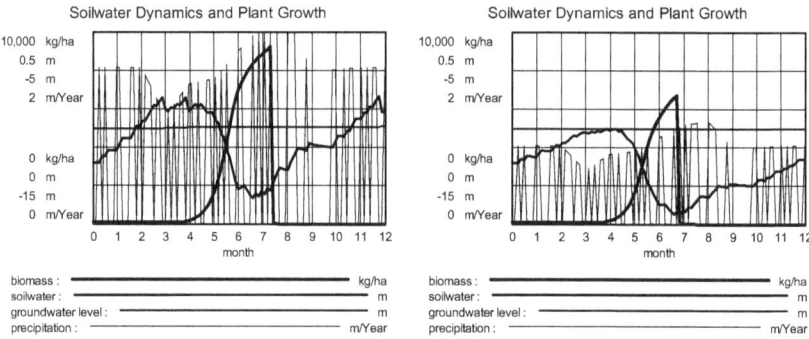

Figure Z310c: Soil water and plant growth for humus rich soil in a normal year.
Figure Z310d: Soil water and plant growth for humus rich soil in a dry year.

Figure Z310d shows results for a dry year (annual precipitation is halved by MULTIPLIER FOR RAINFALL = 0.5) under otherwise identical conditions. The *biomass* at HARVEST DATE is reduced to about 60 percent of MAX BIOMASS because of continuously insufficient supply of *soil water*. There is no *percolation*, and since the *ground water level* cannot be reached by *capillary rise* and an (underground) inflow or drain of ground water is not assumed in the model, the *ground water level* remains constant.

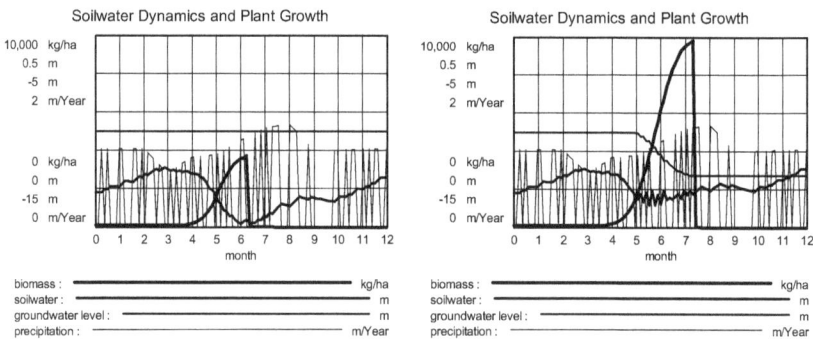

Figure Z310e: Plant growth in a dry year for unfavorable soil conditions.
Figure Z310f: Plant growth for unfavorable soil conditions, but with irrigation.

Figure Z310e shows results for the same weather conditions (MULTIPLIER FOR RAINFALL = 0.5) and unfavorable soil parameters. In this simulation the following parameters were changed: DEPTH OF PLOW LAYER = 0.1, ORGANIC MATTER IN SOIL = 0.01, SOIL COMPACTION PARAMETER = 1, SOIL COVER = 0; remaining parameters as before. Under these conditions the *biomass* grows to less than 40 percent of MAX BIOMASS, which amounts to complete harvest loss.

Figure Z310f shows that even under these difficult soil conditions high crop yield can be achieved by irrigation with ground water (PUMP RATE = 150) if soil moisture becomes too low. In this example it leads to lowering of the *ground water level* by about 4 m (no inflow from elsewhere).

Because of the simplicity of the model these results can only claim limited precision but they already show the significance of soil cover and of humus content of the topsoil for the maintenance of sufficient soil moisture despite limited or little irrigation. The time plots furthermore provide some insight into the seasonal changes of soil humidity, irrigation needs, and ground water level as function of soil parameters, rainfall, and field management.

Exercises

1. Investigate systematically, under otherwise equal conditions, the influence of the different soil parameters on water availability under marginal conditions (drought year, rare periods of rainfall = long WEATHER PERIOD). What additional effects can be obtained by foil cover (reduction of evaporation)? What consequences does this have for irrigation demand?

2. While retaining the other parameters, investigate the influence of soil texture (CLAY FRACTION OF SOIL) on water availability, water demand, and harvest yield.

3. Apply randomized precipitation and relatively low annual precipitation value (*rain amount* after 1 year) of e.g. 500 mm/year over several years. Change for every year the initial number ("seed") of the randomizer (third argument in RANDOM UNIFORM in the equation for the variable "*new value*"). What is your "experience" with your "harvests" over a span of 5 or 10 years for (a) humus-rich clay soil, and (b) humus-poor sandy soil (LOW CLAY FRACTION SOIL)? In case you irrigate: What differences appear with respect to *irrigation amount* and *ground water level*? Under what conditions is it possible to maintain a constant (average) ground water level despite irrigation?

References

Ruhrstickstoff AG 1983: *Faustzahlen für Landwirtschaft und Gartenbau*. Landwirtschaftsverlag Münster-Hiltrup.

Scheffer/Schachtschabel 1979: *Lehrbuch der Bodenkunde*. 10. Auflage, Enke, Stuttgart.

Finck, A. 1982: *Pflanzenernährung in Stichworten*. Hirt, Kiel, S. 75-81.

Z311 Nutrient dynamics

Simulation task

Several processes provide to the soil the nutrient nitrogen which is essential for plant growth: atmospheric precipitation, nitrogen fixation by free-living bacteria and by plants of the legume family (peas, clover), ammonium and nitrate from decomposition of organic matter in the soil, and fertilization with mineral and organic fertilizers. Nitrogen is removed from the soil by nutrient uptake of plants, by leaching, and by denitrification etc. Each of these processes has different characteristics, and in particular a different characteristic time constant. Together these processes represent a complex dynamic system. Since the time constants are of the order of days for some subprocesses (nutrient uptake by plants, fertilization effects) or even of years (decomposition of organic matter), relatively complex dynamic behavior can be expected which will not show up in an annual balance of nitrogen flows. However, a simulation model of this dynamic system can provide a more detailed description. In particular where nitrogen availability must essentially rely on soil processes with their large time constants, as in organic (ecological) agriculture, simulations of nutrient dynamics can provide useful insights. Important applications are: scheduling of fertilization (type, amount, date), preparation of crop rotation schemes, and farm management decision support.

The model presented here is a dynamic simulation model of the most important processes of the soil nitrogen system and its interactions with the processes of plant growth and decomposition of organic matter. The state variables are: plant biomass, plant available nitrogen, carbon and nitrogen in undecomposed organic matter (nutrient humus), carbon in permanent humus. Mathematically the system is therefore described by a system of five nonlinear ordinary differential equations.

It was aim of the model development to capture basic dynamic processes and behaviors of the system in a model system as small as possible. It has a variety of adjustable parameters for soil conditions, fertilization, and different species of crop plants. It can therefore be used for a wide spectrum of conditions to get a better understanding of the real system by experimenting with the model system. This better "feel" for the system should enable the model user to make more efficient management decisions with regard to the use of scarce resources also in the real system. Because of its compactness and limited database one should not expect from the model precise management information for well-defined conditions. To accomplish this task reliably, the model would have to be expanded considerably. Although the model can be operated in isolation, it has however been designed for linkage with model Z310 "Soil water dynamics" discussed above.

Simulation model

The structure of the model system is shown in the simulation diagrams of Figures Z311a and b. The corresponding model equations are listed in the following. The model computes the growth of a field crop up to harvest as function of plant-available nitrogen for a variety of soil, fertilization and plant parameters. Plant-available nitrogen is determined from fertilizer input, leaching losses, and in particular from decomposition processes and losses in nutrient humus and permanent humus, which are a function of the ratio of carbon to nitrogen (C/N ratio). Crop residues, and plant residues remaining in the soil and in the field from previous cultivation periods are decomposed together with any manure and/or compost added, and are transformed into permanent humus and plant-available nitrogen. These processes are a function of C/N ratio and seasonal soil temperature.

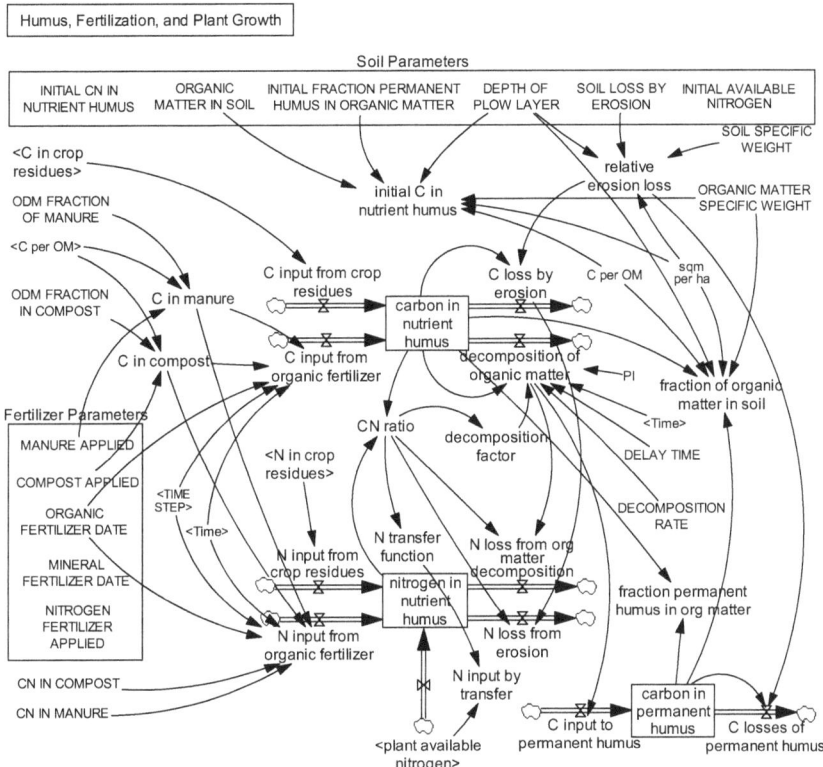

Figure Z311a: Simulation diagram for soil nutrient dynamics, part 1.

The state variables of the model system are: carbon in easily decomposable organic matter (*carbon in nutrient humus*), *nitrogen in nutrient humus*, *carbon in permanent humus*, *plant available nitrogen* and *relative plant biomass* of the field crop.

At the beginning of the simulation (usually the start of a new year) the initial amount of *carbon in nutrient humus* is determined from the amount of ORGANIC MATTER IN SOIL, DEPTH OF PLOW LAYER and INITIAL FRACTION PERMANENT HUMUS IN ORGANIC MATTER (OM = organic matter, ODM = organic dry matter). The initial *nitrogen in nutrient humus* is determined from *initial C in nutrient humus* by applying the initial carbon/nitrogen relationship (INITIAL CN IN NUTRIENT HUMUS). The initial value for *carbon in permanent humus* follows from INITIAL FRACTION PERMANENT HUMUS IN ORGANIC MATTER. The initial value for *plant available nitrogen* follows from INI-

TIAL AVAILABLE NITROGEN. The initial *relative plant biomass* corresponds to the relative amount of seed (usually approximately 2 percent of maximum biomass).

The state variable *carbon in nutrient humus* depends on the rates *decomposition of organic matter, C loss by erosion, C input from crop residues,* and *C input from organic fertilizer.* In a similar way the state of *nitrogen in nutrient humus* is determined by the rates *N loss from org matter* decomposition, *N loss from erosion, N input from crop residues, N input from organic fertilizer* and *N input by transfer* from the *plant available nitrogen* pool. The latter effect occurs if the *CN ratio* is too wide, and soil organisms must temporarily "borrow" nitrogen from the soil to fulfill their functions. The *N transfer function* therefore depends on the current *CN ratio.* The decomposition rate has a maximum for a C/N ratio of about 20. In addition, *decomposition of organic matter* depends on the amount of organic matter present (expressed by *carbon in nutrient humus*) and on soil temperature which is primarily a function of the solar radiation and therefore of seasonal *time.*

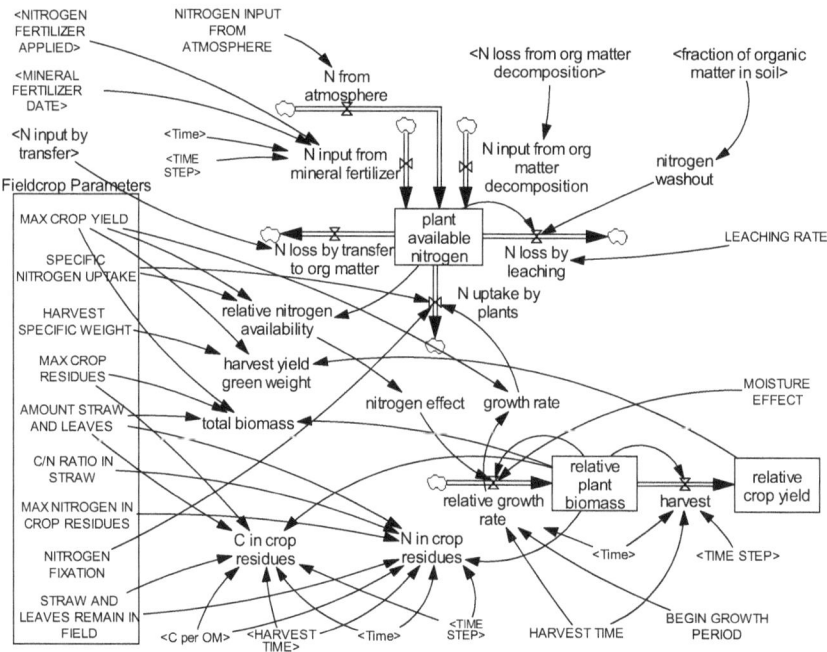

Figure Z311b: Simulation diagram for soil nutrient dynamics, part 2.

The amount of *plant available nitrogen* increases with *N input from org matter decomposition, input of N from atmosphere* and *N input from mineral fertilizer* and is reduced according to *N uptake by plants, N loss by transfer to org matter,* and *N loss by leaching. N loss by leaching* decreases if the *fraction of organic matter in soil* is higher. The pool of *carbon in permanent humus* suffers *C losses of permanent humus* from gradual decomposition and from erosion. Only a small portion (approximately

25 percent) of *carbon in nutrient humus* is transferred as *C input to permanent humus*; the remainder escapes into the atmosphere as carbon dioxide.

The current state of *relative plant biomass* depends on *relative growth rate,* which in turn is an empirical function (*nitrogen effect*) of *relative nitrogen availability* in the soil. Growth starts at BEGIN GROWTH PERIOD and ends at HARVEST TIME. It also depends on relative availability of water (MOISTURE EFFECT); this influence is calculated in model Z310 "Soil water dynamics". The respective *N uptake by plants* depends on *growth rate* corresponding to *relative growth rate* and plant specific MAX CROP YIELD. The yield of *total biomass* is determined from *relative plant biomass* and MAX CROP YIELD. Its value at HARVEST TIME is used to determine the plant-specific amounts of straw, leaves and other crop residues remaining in the field after harvest.

The simulation usually covers the span of a year, but it can also be used to compute crop rotation schemes covering several years, and consequent long-term C and N dynamics in the soil. In this case the values of state variables must be noted down at the end of each year and entered as initial values for the next year. In addition, amounts and dates of fertilization as well as plant parameters have to be changed to reflect the new conditions.

Abbreviations: OM = organic matter, ODM = organic dry matter

Soil parameters
INITIAL CN IN NUTRIENT HUMUS = 20 [1]
ORGANIC MATTER IN SOIL = 0.04 [1] *volume fraction*
INITIAL FRACTION PERMANENT HUMUS IN ORGANIC MATTER = 0.75 [1] *volume
    fraction*
DEPTH OF PLOW LAYER = 0.2 [m]
SOIL LOSS BY EROSION = 5000 [kg/(Year*ha)]
INITIAL AVAILABLE NITROGEN = 50 [kg/ha]
LEACHING RATE = 0.3 [1/Year]
SOIL SPECIFIC WEIGHT = 1600 [kg/(m*m*m)]

Fertilization parameters
MANURE APPLIED = 15000 [kg/ha]
COMPOST APPLIED = 10000 [kg/ha]
ORGANIC FERTILIZER DATE = 0.2 [Year] *0.20 = 10th week = beginning of March*
MINERAL FERTILIZER DATE = 0.25 [Year] *0.25 = 13th week = end of March*
NITROGEN FERTILIZER APPLIED = 80 [kg/ha]
CN IN COMPOST = 15 *1 [1] *ratio of C to N in compost*
CN IN MANURE = 20 *1 [1] *ratio of C to N in manure*
ODM FRACTION OF MANURE = 0.25 *1 [1] *kg ODM per kg fresh matter*
ODM FRACTION IN COMPOST = 0.35 *1 [1] *kg ODM per kg fresh matter*
NITROGEN INPUT FROM ATMOSPHERE = 25 [kg/(ha*Year)]
DECOMPOSITION RATE = 0.2 [1/Year]

Crop parameters
MAX CROP YIELD = 6500 [kg/ha]
SPECIFIC NITROGEN UPTAKE = 0.029 [1] *kg N per kg ODM harvest yield*
HARVEST SPECIFIC WEIGHT = 1.15 [1] *kg fresh weight per kg ODM*
MAX CROP RESIDUES = 1700 [kg/ha]
AMOUNT STRAW AND LEAVES = 7700 [kg/ha]

C/N RATIO IN STRAW = 80 [1]
MAX NITROGEN IN CROP RESIDUES = 17 [kg/ha]
NITROGEN FIXATION = 0 [1] *legumes only, otherwise = 0, in kg N per kg ODM yield*
STRAW AND LEAVES REMAIN IN FIELD = 1 [1] *straw or leaves remain in field = 1,*
     are harvested = 0
BEGIN GROWTH PERIOD = 0.3 [Year]
HARVEST TIME = 0.6 [Year]
MOISTURE EFFECT = 1 [1]

Constants

ORGANIC MATTER SPECIFIC WEIGHT = 650 [kg/(m*m*m)] *spec. weight, dry matter*
PI = 3.14159 *1 [1/Year]
sqm per ha = 10000 *1 [m*m /ha]
DELAY TIME = 0.25 [Year]
C per OM = 0.47 [1] *kg C per kg OM*

Carbon in nutrient humus and permanent humus

initial C in nutrient humus = ORGANIC MATTER IN SOIL *(1 -INITIAL FRACTION
     PERMANENT HUMUS IN ORGANIC MATTER) *DEPTH OF PLOW LAYER
     *sqm per ha *ORGANIC MATTER SPECIFIC WEIGHT *C per OM [kg/ha]
C input from crop residues = C in crop residues [kg/(ha*Year)]
C in manure = MANURE APPLIED *ODM FRACTION OF MANURE *C per OM
     [kg/ha]
C in compost = COMPOST APPLIED *ODM FRACTION IN COMPOST *C per OM
     [kg/ha]
C input from organic fertilizer = IF THEN ELSE (ABS(Time -ORGANIC FERTILIZER
     DATE) < TIME STEP/2, (C in manure +C in compost) /TIME STEP, 0)
     [kg/(Year*ha)]
relative erosion loss = SOIL LOSS BY EROSION /(DEPTH OF PLOW LAYER
     *sqm per ha *SOIL SPECIFIC WEIGHT) [1/Year]
C loss by erosion = relative erosion loss *carbon in nutrient humus [kg/(ha*Year)]
decomposition factor = WITH LOOKUP (1 /CN ratio, ([(0, 0) -(0.05, 1)], (0, 0), (0.01,
     0.05), (0.025, 0.25), (0.033, 0.5), (0.04, 0.9), (0.05, 1))) [1]
decomposition of organic matter = DECOMPOSITION RATE *decomposition factor
     *carbon in nutrient humus *(1 +0.5 *SIN (2 *PI *(Time -DELAY TIME)))
     [kg/(ha*Year)]
carbon in nutrient humus = INTEG (C input from organic fertilizer +C input from crop
     residues -C loss by erosion -decomposition of organic matter, initial C in nutrient
     humus) [kg/ha] *nutrient humus: undecomposed organic matter*
fraction of organic matter in soil = (carbon in nutrient humus +carbon in permanent
     humus) /(DEPTH OF PLOW LAYER *sqm per ha *ORGANIC MATTER
     SPECIFIC WEIGHT *C per OM) [1]
C input to permanent humus = 0.25 *decomposition of organic matter [kg/(ha*Year)]
C losses of permanent humus = carbon in permanent humus *(0.2 +relative erosion
     loss) [kg/(ha*Year)]
carbon in permanent humus = INTEG (C input to permanent humus -C losses of
     permanent humus, (ORGANIC MATTER IN SOIL *(1 -INITIAL FRACTION
     PERMANENT HUMUS IN ORGANIC MATTER) *DEPTH OF PLOW LAYER
     *sqm per ha *ORGANIC MATTER SPECIFIC WEIGHT *C per OM) *INITIAL
     FRACTION PERMANENT HUMUS IN ORGANIC MATTER /(1 -INITIAL
     FRACTION PERMANENT HUMUS IN ORGANIC MATTER)) [kg/ha]

fraction permanent humus in org matter = carbon in permanent humus /(carbon in permanent humus +carbon in nutrient humus) [1]

Nitrogen in nutrient humus
CN ratio = carbon in nutrient humus /nitrogen in nutrient humus [1]
N transfer function = WITH LOOKUP (1 /CN ratio, ([(0, 0) -(0.05, 1)], (0, 1), (0.01, 0.95), (0.025, 0.75), (0.033, 0.5), (0.04, 0.2), (0.05, 0))) [1/Year]
N loss from org matter decomposition = decomposition of organic matter /CN ratio [kg/(ha*Year)]
N loss from erosion = C loss by erosion /CN ratio [kg/(ha*Year)]
N input from crop residues = N in crop residues [kg/(ha*Year)]
N input from organic fertilizer = IF THEN ELSE (ABS (Time -ORGANIC FERTILIZER DATE) < TIME STEP/2, (C in manure /CN IN MANURE +C in compost /CN IN COMPOST) /TIME STEP, 0) [kg/(Year*ha)]
N input by transfer = N transfer function *plant available nitrogen [kg/(ha*Year)]
nitrogen in nutrient humus = INTEG (N input from organic fertilizer +N input from crop residues +N input by transfer +N loss from org matter decomposition -N loss from erosion, initial C in nutrient humus /INITIAL CN IN NUTRIENT HUMUS) [kg/ha]

Plant-available nitrogen
N from atmosphere = NITROGEN INPUT FROM ATMOSPHERE [kg/(Year*ha)]
N input from org matter decomposition = N loss from org matter decomposition [kg/(ha*Year)]
N input from mineral fertilizer = IF THEN ELSE (ABS (Time -MINERAL FERTILIZER DATE) < TIME STEP /2, NITROGEN FERTILIZER APPLIED /TIME STEP, 0) [kg/(ha*Year)]
nitrogen leaching = WITH LOOKUP (fraction of organic matter in soil, ([(0, 0) -(1, 1)], (0, 1), (0.05, 0.5), (0.1, 0.2), (1, 0.1))) [1]
N loss by leaching = LEACHING RATE *plant available nitrogen *nitrogen leaching [kg/(ha*Year)]
N loss by transfer to org matter = N input by transfer [kg/(ha*Year)]
N uptake by plants = growth rate *(SPECIFIC NITROGEN UPTAKE -NITROGEN FIXATION) [kg/(ha*Year)]
plant available nitrogen = INTEG (N from atmosphere +N input from org matter decomposition +N input from mineral fertilizer -N uptake by plants -N loss by leaching -N loss by transfer to org matter, INITIAL AVAILABLE NITROGEN) [kg/ha]
relative nitrogen availability = plant available nitrogen /(MAX CROP YIELD *SPECIFIC NITROGEN UPTAKE) [1]

Plant growth
nitrogen effect = WITH LOOKUP (relative nitrogen availability, ([(0, 0) -(10, 1)], (0, 0), (0.2, 0.2), (0.35, 0.5), (0.5, 0.8), (1, 1), (2, 0.9), (3, 0.4), (5, 0.1), (10, 0))) [1]
relative growth rate = IF THEN ELSE ((Time -INTEGER(Time) < BEGIN GROWTH PERIOD) :OR: (Time −INTEGER (Time) > HARVEST TIME), 0, (25*(20/52) /(HARVEST TIME -BEGIN GROWTH PERIOD)) *relative plant biomass *(1 -relative plant biomass) *MOISTURE EFFECT *nitrogen effect) [1/Year]
growth rate = relative growth rate *MAX CROP YIELD [kg/(ha*Year)]
harvest = IF THEN ELSE (ABS (Time -HARVEST TIME) < TIME STEP/2, relative plant biomass /TIME STEP, 0) [1/Year]

harvest yield green weight = HARVEST SPECIFIC WEIGHT *relative crop yield
    *MAX CROP YIELD [kg/ha] *Frischgewicht*
relative plant biomass = INTEG (relative growth rate -harvest, 0.01) [1]
relative crop yield = INTEG (harvest, 0) [1]
C in crop residues = IF THEN ELSE (ABS(Time -HARVEST TIME) < TIME STEP/2,
    (C per OM *relative plant biomass *(STRAW AND LEAVES REMAIN IN FIELD
    *AMOUNT STRAW AND LEAVES +MAX CROP RESIDUES)) /TIME STEP, 0)
    [kg/(ha*Year)]
N in crop residues = IF THEN ELSE (ABS (Time -HARVEST TIME) < TIME STEP/2,
    (C per OM *relative plant biomass *(STRAW AND LEAVES REMAIN IN FIELD
    *AMOUNT STRAW AND LEAVES /C/N RATIO IN STRAW +MAX NITROGEN IN
    CROP RESIDUES)) /TIME STEP, 0) [kg/(ha*Year)]
total biomass = relative plant biomass *(MAX CROP YIELD +AMOUNT STRAW AND
    LEAVES +MAX CROP RESIDUES) [kg/ha]

Simulation time parameters
INITIAL TIME = 0 [Year]
FINAL TIME = 1 [Year]
TIME STEP = 0.01 [Year]

Data for different crops

(see model parameter listings for respective units)

	cereal crops	maize	potatoes	beets	beans, peas	rape	clover, alfalfa	grass
MAX CROP YIELD	6500	7800	9000	9800	4400	3600	7000	5200
SPECIFIC NITRO-GEN UPTAKE	0.029	0.032	0.031	0.035	0.075	0.06	0.029	0.023
HARVEST SPECIFIC WEIGHT	1.15	1.15	5.56	7.14	1.14	1.11	1.14	1.15
AMOUNT STRAW AND LEAVES	7700	8600	6000	6000	6000	6900	0	0
MAX CROP RESI-DUES	1700	2000	1000	1000	2300	1300	4000	4000
C/N RATIO IN STRAW	80	55	30	30	15	30	15	50
MAX NITROGEN IN CROP RESIDUES	17	20	25	25	53	30	75	30
NITROGEN FIXA-TION	0	0	0	0	0.068	0	0.025	0

Simulation results

Because of the large number of parameters which can be changed by the user, this
model can be used for a variety of different conditions. Simulations with this model
assume sufficient water availability (MOISTURE EFFECT = 1). Plant growth for limited
water availability can be computed by coupling the model to model Z310 "Soil water
dynamics"; this is done in model Z312 "Field crop cultivation".

Figure Z311c shows the development for the default parameter set. In this case a cereal crop is grown; the corresponding parameters of the table above are also found in the model equations. At *time* = 0.2 (approximately the beginning of March) the field is fertilized with a large amount of MANURE APPLIED and COMPOST APPLIED. In addition at MINERAL FERTILIZER DATE (*time*) = 0.25 (end of March) NITROGEN FER-TILIZER is APPLIED. *Plant available nitrogen* rises drastically with the mineral fertil-izer input; it is used up quickly however, particularly during the most intensive growth period of the crop. Since decomposition of *organic matter in soil* (manure, compost, crop residues) is a slow process, it leads to only gradual increase of *plant-available nitrogen*, filling this pool slowly again after the harvest.

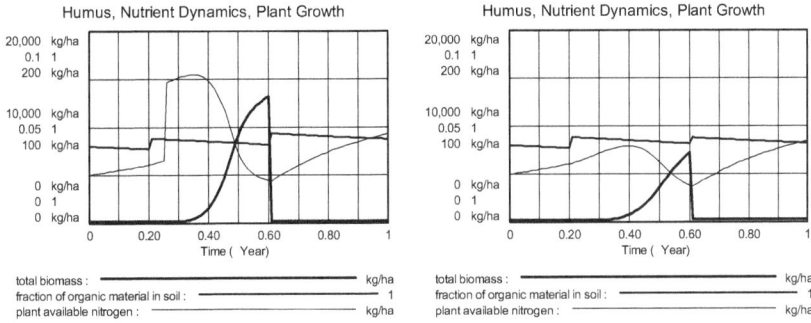

Figure Z311c: Plant growth for organic and mineral fertilization.
Figure Z311d: Plant growth without any fertilization.

Figure Z311d clarifies the considerable influence which nitrogen fertilization (with organic matter or mineral fertilizer) has on plant growth. In this case initial con-ditions and crop type are identical to the previous case except there is no fertilization. Since the chosen soil parameters assume a relatively high initial fraction of ORGANIC MATTER IN SOIL (0.04 = 4 percent), its decomposition initially produces an increase in *plant available nitrogen*, while *N uptake by plants* during the growing season drains this pool, and another increase occurs after harvest. However, the nitrogen supply is not sufficient by far; *total biomass* of the crop only reaches about 40 percent of its potential maximum.

Exercises

1. Experiment with different soil conditions, field crops, crop rotation schemes, fertil-izer applications etc. and try to manage your "farm" over several (simulated) years while achieving high yields. How does soil quality change during this period (*fraction of organic matter in soil* and *plant available nitrogen*)? Calculate your costs and prof-its by applying appropriate prices.
2. In organic farming a livestock density of approximately one cattle unit (CU: 1 cow, or 3 pigs, or 10 sheep, or 50 chickens) per hectare is recommended. Use data from agriculture textbooks to determine what inputs of organic matter and nitrogen in ma-nure can be supplied with this livestock density. What average grain yields can be

expected without additional supply of mineral fertilizer or cultivation of legumes? What grain yields could be achieved in a crop rotation scheme with legumes?

3. Develop your own crop rotation schemes and fertilization schedules and compute achievable yields. Give reasons for your choice, and explain the results.

References

Ruhrstickstoff AG 1983: *Faustzahlen für Landwirtschaft und Gartenbau*. Landwirtschaftsverlag Münster-Hiltrup.

Finck, A. 1982: *Pflanzenernährung in Stichworten*. Hirt, Kiel.

Larcher, W. 1980: *Ökologie der Pflanzen*. UTB/Ulmer, Stuttgart.

Thompson, L.M., Troeh, F.R. 1978: *Soils and Soil Fertility*, 4th ed. McGrawHill, New York.

Scheffer/Schachtschabel 1979: *Lehrbuch der Bodenkunde*, 10. Auflage. Enke, Stuttgart.

Wild, E. (ed.) 1988: *Russell's Soil Condition and Plant Growth*. 11th ed. Longman Group, Harlow UK

Z312 Field crop cultivation

Simulation task

Neither analysis of water availability alone (as in model Z310 "Soil water dynamics") nor of nitrogen availability (as in model Z311 "Nutrient dynamics") is sufficient to compute the dynamics of plant growth. These two models were therefore designed from the start to be coupled to each other. Model Z312 "Field crop cultivation" is the result of this coupling. It allows relatively realistic investigations of field crop production for different crops as function of soil properties, weather, fertilization etc.

Simulation model

To create model Z312 "Field crop cultivation" the two models Z310 "Soil water dynamics" and Z311 "Nutrient dynamics" are linked without changes of structure or content. Only a very few quantities have to be modified or renamed (cf. the following model instructions). The simple model of plant growth contained in model Z310 is replaced by the corresponding model from Z311, which was modified insignificantly.

The coupling process can be carried out very simply (e.g. with the VensimPLE program system) on the PC screen by marking, copying, and then pasting model Z310 next to the simulation diagram of model Z311. The two parts of the complete model Z312 are then coupled according to the model structure shown in Figures Z312a and b and the following modified model instructions.

The model parts were fully explained in the model descriptions for Z310 "Soil water dynamics" and Z311 "Nutrient dynamics".

Necessary program changes:

MAX CROP YIELD = 6500 [kg/ha]
    (replaces 'MAX BIOMASS' from Z310)
moisture effect = WITH LOOKUP (soil moisture, ([(0,0) -(2, 2)], (0, 0), (0.1, 0.2), (0.3, 0.5), (0.5, 1), (1, 1), (2, 1))) [1]
    (replaces 'moisture effect' from Z311)
total biomass = relative plant biomass *(MAX CROP YIELD +AMOUNT STRAW AND LEAVES +MAX CROP RESIDUES) [kg/ha]
    (replaces 'biomass' from Z310)
relative growth rate = IF THEN ELSE((Time –INTEGER (Time) < BEGIN GROWTH PERIOD) :OR: (Time –INTEGER (Time) > HARVEST DATE), 0, (25*(20/52) /(HARVEST DATE - BEGIN GROWTH PERIOD)) *relative plant biomass *(1 -relative plant biomass) *moisture effect * nitrogen effect) [1/Year]
    (replaces 'relative growth' from Z310)
growth rate = relative growth rate *MAX CROP YIELD [kg/(ha*Year)]
    (replaces 'growth rate' from Z310 or Z311)
growth rate total biomass = relative growth rate *total biomass [kg/(Year*ha)]
    (additional)
relative plant biomass = INTEG (relative growth rate -harvest -withering, 0.01) [1]
    (replaces 'relative plant biomass' from Z311)
transpiration = (1/ FERTILIZATION FACTOR) * SPECIFIC TRANSPIRATION RATE *growth rate total biomass /10000 [m/Year]
    (replaces 'transpiration' from Z310)

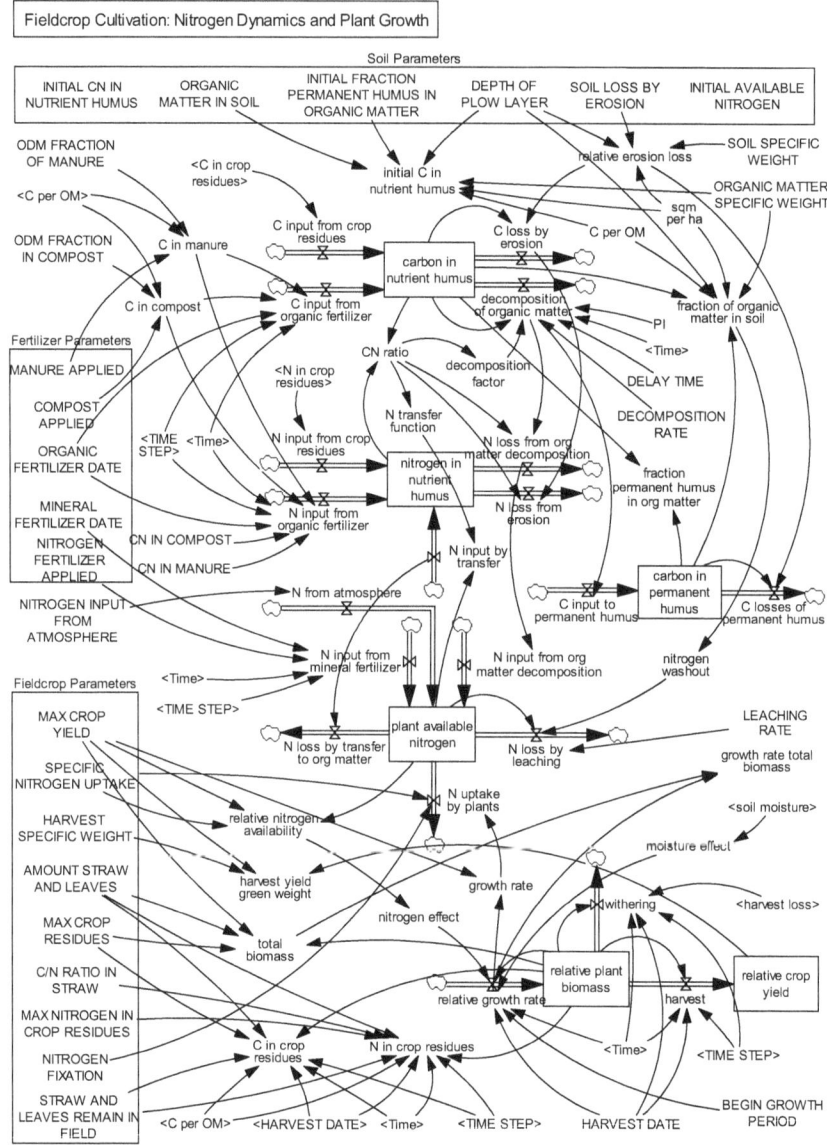

Figure Z312a: Simulation diagram for field crop production – part 1.

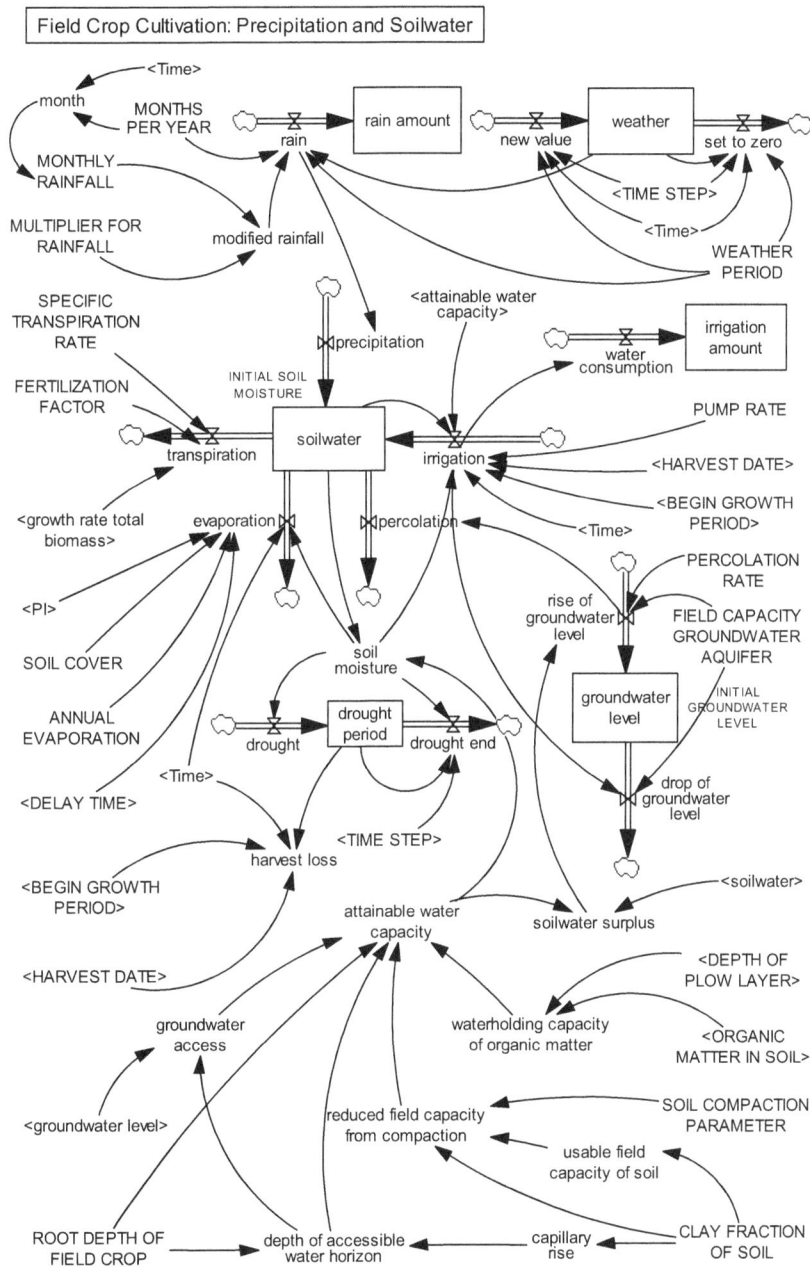

Figure Z312b: Simulation diagram for field crop production – part 2.

Simulation results

Exemplary simulation results are shown in Figure Z312c and d for favorable and un-favorable growth conditions. In both cases randomized precipitation was applied (as in Z310) with a WEATHER PERIOD of 10 days.

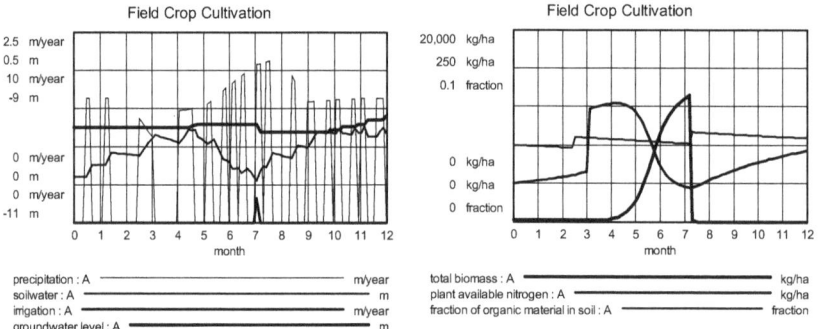

Figure Z312c: Growth under favorable conditions. On the left: water supply; on the right: plant growth, plant available nitrogen, and soil organic matter.

Figure Z312c shows the development for the default parameter setting (heavy organic and mineral fertilization as in Z311: MANURE APPLIED = 15'000 kg/ha and COMPOST APPLIED = 10'000 kg/ha at ORGANIC FERTILIZER DATE = 0.2, NITROGEN FERTILIZER applied = 80 kg/ha at MINERAL FERTILIZER DATE = 0.25; humus rich deep topsoil with 4 percent ORGANIC MATTER IN SOIL and DEPTH OF PLOW LAYER = 0.2 m; normal (Central European) annual precipitation). These favorable conditions are com-parable with those in Figure Z311c and show an almost identical result with high yield, accumulation of *plant available nitrogen*, and slight *ground water* increase over the span of a year.

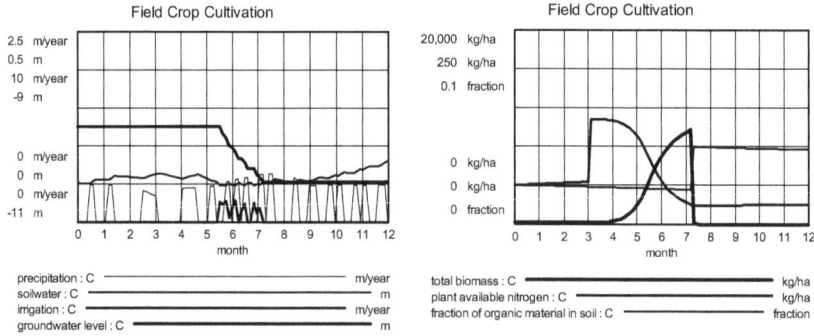

Figure Z312c: Growth under unfavorable conditions. On the left: Water supply; on the right: Plant growth, plant available nitrogen, and soil organic matter.

Figure Z312d shows the development for poor soil conditions, low precipitation and purely mineral nitrogen fertilization (MANURE APPLIED = 0, COMPOST APPLIED = 0, mineral NITROGEN FERTILIZER APPLIED = 80 at MINERAL FERTILIZER DATE = 0.25; humus-poor thin topsoil with 2 percent ORGANIC MATERIAL IN SOIL and DEPTH OF PLOW LAYER = 0.04 m; MULTIPLIER FOR RAINFALL = 0.3). In this case yield is reduced considerably. The necessary irrigation during the growth period leads to a drop of ground water level of about 0.6 m.

Some conclusions

Simulation models representing with reasonable accuracy the actual and interconnected processes of plant growth and of nutrient and water dynamics in the soil can be used to obtain conclusions about the dynamics of field crop cultivation under a variety of different conditions. From experience with such models several conclusions can be drawn:

- Plant cultivation is characterized by dynamic processes with periods of days (e.g. pest attack), months (e.g. plant growth), and years (e.g. erosion and humus depletion).
- To control and optimize these processes, the farmer must intervene at the right time with the right choice and amount of appropriate measures.
- Since this task requires cybernetic and systemic knowledge, it can in principle be supported by system models.
- Agricultural system models can already lead to better decisions merely by facilitating better understanding of the plant production system.
- More detailed models can be used as decision aids for concrete local field management.
- At the operational level of the farm, the short-term success of production is primarily dependent on structure of the operation and its use of the means of production, but in the long run it depends decisively on conservation of soil fertility, and thus on soil organic matter.
- In organic agriculture a well-balanced ratio of livestock per area, recycling of animal and plant wastes and residues, and the use of "green manure" (in particular: legumes) can lead to high levels of humus, nitrogen, and soil moisture, and sustainable high yields (comparable to traditional intensive agriculture) even if no mineral fertilization is used.
- In addition to the physical characteristics of the soil, the organic matter content of topsoil is primarily decisive for soil water availability. The harvest yield in dry years in particular depends on this factor.
- Decomposition processes and the supply of organic matter (crop residues, manure, humus) play a decisive role for the supply of plant available nitrogen in the soil.
- High nitrogen inputs can be achieved by application of "green manure" and by nitrogen-fixing legumes (clover, alfalfa, peas, beans, acacia trees etc.).
- A carefully balanced crop rotation scheme is decisive for supplying adequate soil nitrogen to achieve high crop yields. Simulation models can assist in the development of successful crop rotation schemes.

Exercises

1. Couple submodels Z310 and Z311 as described and make sure that model Z312 "Field crop cultivation" works correctly and reproduces the results shown.

2. Simulate the cultivation of different plants (cf. data table in Z311 "Nutrient dynamics") for different fertilization, soil, and precipitation conditions, and try to understand the dynamic development of key variables (primarily the state variables) and to develop a "feel" of the system.

(*Hint:* Before running a simulation, consider what developments can be expected, and then compare your expectations with actual simulation results.)

3. In organic agriculture (where mineral nitrogen fertilizer is not used) ingenious crop rotation schemes are extremely important for ensuring sufficient water and nutrient supplies during the plant growth period. Try to develop advantageous crop rotation schemes with the help of the model. For this task you must use the model in such a way (or modify it accordingly) that state variable values at the end of one year are used as initial values at the start of the next year. Start with the initial values of the default parameter setting. In this way, simulate the following seven year crop rotation scheme (which is typical for organic agriculture). Use NITROGEN FERTILIZER APPLIED = 0, SOIL COVER = 1, crop data from Table in Z311. Pay special attention to the time plot of *plant available nitrogen.* How close are the crop yields to MAX CROP YIELD of the respective crop?

year	1	2	3	4	5	6	7
crop choice	clover	clover	potatoes	wheat	beans	wheat	barley
manure (t/hectare):	17	17	44	13.3	6.7	6.7	6.7
straw remains in field:	0	0	1	0	1	0	0

4. With what restrictions regarding the amount of precipitation (MULTIPLIER FOR RAINFALL) can this cultivation system cope without irrigation and without dramatic drops in yield (PUMP RATE = 0)?

5. Simulate the long-term development (also 7 years) of a grain monoculture which is fertilized merely with mineral nitrogen fertilizer (and no organic matter inputs). How do the characteristic measures of soil quality (*fraction of organic matter in soil, carbon in permanent humus, attainable water capacity* and others) develop in this case? Which inputs of mineral fertilizer are necessary, and which lowering of the ground water level occurs if a high crop yield is to be achieved by irrigation and mineral fertilization under conditions of reduced precipitation (as in Problem 4)?

References

Cf. models Z310 "Soil water dynamics" and Z311 "Nutrient dynamics"

Bossel, H. 1985: *Umweltdynamik – 30 Programme für kybernetische Umwelterfahrungen.* TeWi Verlag, München (275-315).

Bossel, H. et al. 1988: *Dokumentation zum Fruchtfolgeberatungssystem FELDSIM.* Gesamthochschule Kassel, Wiss. Zentrum Mensch, Umwelt, Technik, AG Ressourcen- und Systemforschung, RSF 88-3.

France, J., Thornley, J. H. M. 1984: *Mathematical Models in Agriculture*. Butterworths, London.

Penning de Vries, F. W. T., van Laar, H. H. 1982: *Simulation of plant growth and crop production*. Pudoc, Wageningen.

Penning de Vries, F. W. T. et al. 1989: *Simulation of ecophysiological processes of growth in several annual crops*. Pudoc, Wageningen.

Richter, O. 1985: *Simulation des Verhaltens ökologischer Systeme*. VCH, Weinheim.

Z313 Food production

Simulation task

A normal diet requires an energy input of about 10'000 kilojoules (kJ) per day for an adult (1 kJ = 0.239 kcal). This should contain at least 40 g of proteins from animal and/or vegetable sources. Even if this protein requirement would be fully covered by animal protein, the share of animal-derived food would only amount to about 10 percent of organic dry matter. In industrial countries today this share is significantly higher; it amounts to about 40 percent in Germany.

Food from animal sources originates at a trophic level at least one level above food from vegetable sources. Feed and fodder for animals must also be produced at this lower (plant) level. About 10 calories at the plant level are required to produce 1 calorie at the animal (or human) level. This means that the area needed to produce food from animal sources is at least ten times as large as that for food from vegetable sources having the same nutritional value. Expressed differently: From the same area of agricultural land about 10 times as many people can be fed on a vegetarian diet than on a diet based solely on animal sources. The larger the share of animal products in the diet, the larger the area required for the production of food. Obviously this simple fact has considerable significance for the world food supply.

If national eating habits shift from animal-based food to vegetable-based food or *vice versa*, this will have significant consequences for the structure of agriculture, for livestock numbers, for the ratio of livestock feed to grain for human consumption, for imports and exports etc. The model is designed to help clarify these relationships and particularly the dynamic changes resulting from changing diet composition in one or the other direction.

Simulation model

The model computes population numbers and food demand from animal and vegetable sources using a simple population model. The amount of livestock (meat, milk, and egg production) determines whether the demand can be met in the desired ratio of animal vs. vegetable based foods. If consumption goal and production do not agree, then the livestock amount is changed gradually by changing the feed volume. Livestock feed originates (1) from fodder produced on the (fixed) grassland area and (2) feed produced on cultivated land (after covering human food needs). Shortages can be met by imports (surpluses can be exported) or can be made up by expansion of the arable land area (as far as possible).

The simulation diagram of the model is shown in Figure Z313a. It contains four state variables: *population*, the *area of arable land* currently in production, *livestock* and *grain supply* which is used for livestock feed and the human diet (and includes also other vegetable food sources like soy bean, cassava, rape etc.). The model equations are listed completely in the following.

The current *population* is computed from the parameters BIRTH RATE and DEATH RATE and the corresponding *births* and *deaths* using a simple population model. Multiplied by DAILY FOOD DEMAND PER CAP the *population* size yields the *food demand* (expressed in energy units). A part of this demand is met by *food from livestock production* which corresponds to *livestock sales*.

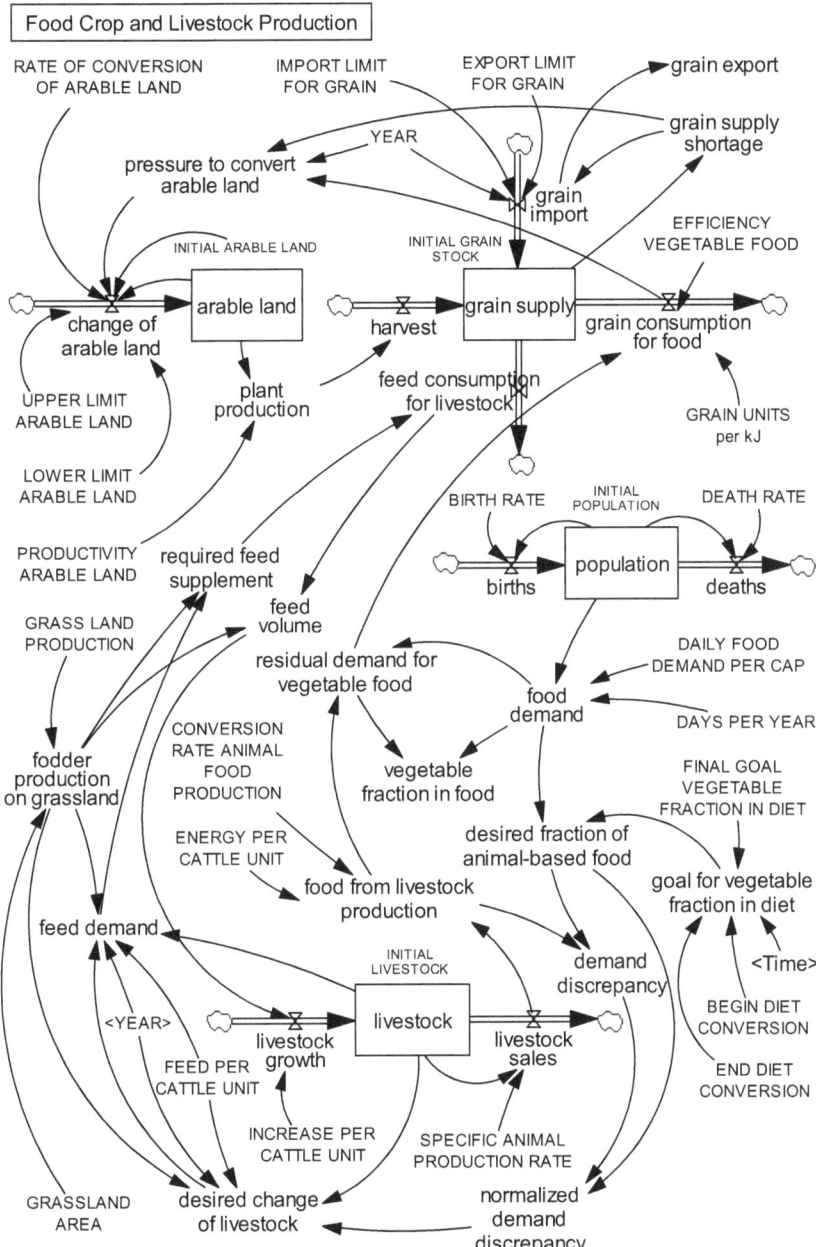

Figure Z313a: Simulation diagram of food supply from animal and vegetable sources.

Livestock growth is a function or *feed volume*. A *demand discrepancy* arises between the *desired fraction of animal-based food* and actual *food from animal production*, which leads to a *desired change of livestock*. The *feed demand* changes correspondingly. Part of the *feed demand* is covered by *fodder production on grassland* (pasture, hay, silage). The remaining *required feed supplement* must be covered by feed from the *grain supply* and is therefore in competition with use for human consumption.

The *grain supply* must primarily be filled by the *harvest* from *plant production*. This corresponds to the currently used area of *arable land* and its PRODUCTIVITY ARABLE LAND. From *grain supply shortage* (or surplus) follows corresponding *grain import* (or export) corresponding to given IMPORT or EXPORT LIMIT FOR GRAIN. If this is not sufficient to compensate for the *grain supply shortage*, *pressure to convert arable land* appears and a corresponding *change of arable land* by fallowing or cultivation occurs.

The numerical values used in the model correspond to statistical data or are conversion factors between different energy units used in statistics. The different variables, their meaning, and their units of measurement are listed in the following.

Conversion factors used in the model
(partly derived from *Faustzahlen* 1983 and *Stat. Jahrbuch* 1982)
GU = grain unit, CU = cattle unit; M = Mega = 10^6
specific feed need per cattle unit = 3.97 tGU/CU (per year)
specific net growth of animals per grain unit =
    = 3.2 CU /tGU * efficiency of animal production
    = 3.2 * 0.27 = 0.89 CU /tGU
efficiency of animal production = 0.27
    = annual production / annual feed input *(meat, milk, eggs weighted)*
specific animal production = 3.48 (CU /a)/ CU *(meat, milk, eggs weighted)*
conversion factor of animal production =
    = amount for consumption / amount produced = 0.68
conversion of cattle unit CU to energy equivalent: $3.5 * 10^6$ kJ/ CU
conversion of kJ to grain units GU: $0.087 * 10^{-3}$ GU/kJ
efficiency of vegetable food consumption =
    = vegetable food demand / amount consumed = 0.92
average production of plowland: 5.97 tGU/ha
average production of grassland: 3.2 tGU/ha

Parameters, constants, initial values and conversions
INITIAL POPULATION = 8e+007 [Person]
BIRTH RATE = 0.009 [1/Year]
DEATH RATE = 0.012 [1/Year]
GRASSLAND AREA = 7.2 [Mha]
GRASS LAND PRODUCTION = 3.2 [MtGU/(Mha *Year)]
INITIAL ARABLE LAND = 10.9 [Mha]
UPPER LIMIT ARABLE LAND = 1 [1] *relative to initial arable land*
LOWER LIMIT ARABLE LAND = 0.8 [1] *relative to initial arable land*
PRODUCTIVITY ARABLE LAND = 5.97 [MtGU/(Mha*Year)]
RATE OF CONVERSION OF ARABLE LAND = 1 [Mha/Year]
FINAL GOAL VEGETABLE FRACTION IN DIET = 0.9 [1] *share of vegetable-based*
    food in total food

BEGIN DIET CONVERSION = 2005 [Year]
END DIET CONVERSION = 2015 [Year]
INITIAL GRAIN SUPPLY = -10 [MtGU]
IMPORT LIMIT FOR GRAIN = 15 [MtGU/Year]
EXPORT LIMIT FOR GRAIN = 20 [MtGU/Year]
DAILY FOOD DEMAND PER CAP = 12500 [kJ/(Person*Day)]
GRAIN UNITS per kJ = 8.7e-014 [MtGU/kJ]
EFFICIENCY VEGETABLE FOOD = 0.92 [1]
INITIAL LIVESTOCK = 17.5 [MGVE]
CONVERSION RATE ANIMAL FOOD PRODUCTION = 0.68 [1]
SPECIFIC ANIMAL PRODUCTION RATE = 3.48 [1/Year]
INCREASE PER CATTLE UNIT = 0.89 [MCU /MtGU]
ENERGY PER CATTLE UNIT = 3.5e+012 [kJ/MCU]
FEED PER CATTLE UNIT = 3.97 [MtGE/(MCU *Year)]
YEAR = 1 [Year]
DAYS PER YEAR = 365 [Day/Year]

Dynamics
births = BIRTH RATE *population [Person/Year]
deaths = DEATH RATE *population [Person/Year]
population = INTEG (births -deaths, INITIAL POPULATION) [Person]
food demand = DAYS PER YEAR *DAILY FOOD DEMAND PER CAP *population
    [kJ/Year]
goal for vegetable fraction in diet = IF THEN ELSE (Time <= BEGIN DIET CONVER-
    SION, 0.6, IF THEN ELSE (Time >= END DIET CONVERSION, FINAL GOAL
    VEGETABLE FRACTION IN DIET, (0.6 +(Time -BEGIN DIET CONVERSION)
    *(FINAL GOAL VEGETABLE FRACTION IN DIET -0.6) /(END DIET CONVER-
    SION -BEGIN DIET CONVERSION)))) [1]
desired fraction of animal-based food = (1 -goal for vegetable fraction in diet)
    *food demand [kJ/Year]
residual demand for vegetable food = food demand -food from livestock production
    [kJ/Year]
vegetable fraction in food = residual demand for vegetable food /food demand [1]
grain consumption for food = GRAIN UNITS per kJ /EFFICIENCY VEGETABLE
    FOOD *residual demand for vegetable food [MtGU/Year]
grain import = IF THEN ELSE (grain supply shortage /YEAR > IMPORT LIMIT FOR
    GRAIN, IMPORT LIMIT FOR GRAIN, IF THEN ELSE(-grain supply shortage
    /YEAR > (EXPORT LIMIT FOR GRAIN), (-EXPORT LIMIT FOR GRAIN),
    grain supply shortage/YEAR)) [MtGU/Year]
grain export = -grain import [MtGU/Year]
grain supply = INTEG (harvest +grain import -feed consumption for livestock
    -grain consumption for food, INITIAL GRAIN SUPPLY) [MtGE]
grain supply shortage = -grain supply [MtGU]
pressure to convert arable land = (grain supply shortage /YEAR) /grain consumption
    for food [1]
change of arable land = IF THEN ELSE (((pressure to convert arable land > 0)
    :AND: (arable land >= UPPER LIMIT ARABLE LAND *INITIAL ARABLE LAND))
    :OR: ((pressure to convert arable land < 0) :AND: (arable land <= LOWER LIMIT
    ARABLE LAND *INITIAL ARABLE LAND)), 0, RATE OF CONVERSION OF
    ARABLE LAND *pressure to convert arable land) [Mha /Year]
arable land = INTEG (change of arable land, INITIAL ARABLE LAND) [Mha]
plant production = PRODUCTIVITY ARABLE LAND *arable land [MtGE/Year]

harvest = plant production [MtGU/Year]

fodder production on grassland = GRASS LAND PRODUCTION *GRASSLAND AREA
[MtGU/Year]

feed demand = IF THEN ELSE (fodder production on grassland <= FEED PER
CATTLE UNIT *(livestock +desired change of livestock *YEAR), FEED PER
CATTLE UNIT *(livestock +desired change of livestock *YEAR), fodder
production on grassland) [MtGU/Year]

required feed supplement = feed demand -fodder production on grassland
[MtGU/Year]

feed consumption for livestock = IF THEN ELSE (required feed supplement <= 0, 0,
required feed supplement) [MtGU/Year]

feed volume = feed consumption for livestock +fodder production on grassland
[MtGU/Year]

livestock growth = INCREASE PER CATTLE UNIT *feed volume [MCU/Year]

livestock = INTEG (livestock growth -livestock sales, INITIAL LIVESTOCK) [MCU]

livestock sales = SPECIFIC ANIMAL PRODUCTION RATE *livestock [MCU/Year]

food from livestock production = CONVERSION RATE ANIMAL FOOD PRODUC-
TION *ENERGY PER CATTLE UNIT *livestock sales [kJ/Year]

demand discrepancy = desired fraction of animal-based food -food from livestock
production [kJ/Year]

normalized demand discrepancy = demand discrepancy /desired fraction of
animal-based food [1]

desired change of livestock = normalized demand discrepancy *(livestock /YEAR)
*((livestock *FEED PER CATTLE UNIT /fodder production on grassland) -1)
[MCU/Year]

Simulation time parameters
INITIAL TIME = 2000 [Year]
FINAL TIME = 2025 [Year]
TIME STEP = 0.125 [Year]

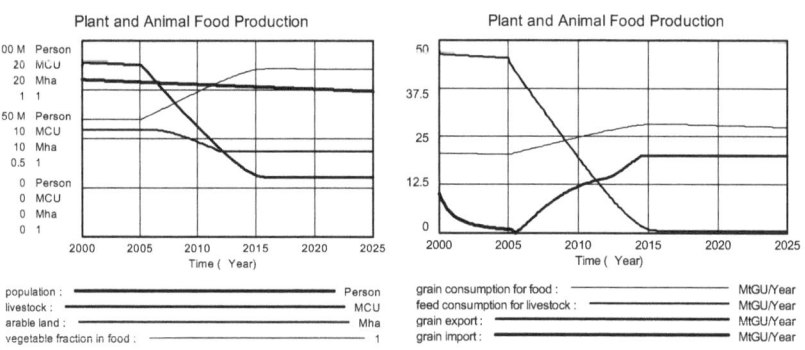

Figure Z313b: Results for a scenario where the vegetarian fraction of the diet increases strongly. Livestock numbers decline strongly, and grain can be exported.

Simulation results

Simulation results are shown in Figure Z313b for a scenario where the VEGETABLE FRACTION IN DIET increases from initially 60 percent to finally 90 percent in relatively short time (from 2005 to 2015) (default parameter setting; approximate data for Germany). For *grain import* (or *grain export*) IMPORT LIMIT FOR GRAIN and EXPORT LIMIT FOR GRAIN were set at 15 and 20 million metric tons of grain units, respectively. The area of *arable land* is not expandable but can be reduced to 80 percent of today's value. The simulation shows a reduction of *livestock* to approximately one third of today's value. The need for cultivation of grain for livestock feed then largely disappears since the feed demand can be met by *fodder production on grass land*. Although the consumption of vegetarian food increases, *grain import* declines quickly so that finally large amounts of grain can be exported although the area of arable land has been reduced by 20 percent. The results clearly show the enormous influence of eating habits on the structure of a country's agricultural production.

Exercises

1. Investigate the consequences of a transition to (a) more, and (b) less meat-based food diet for different assumptions concerning imports and exports and the conversion of arable land.
2. Examine and discuss the influence of the parameter RATE OF CONVERSION OF AR-ABLE LAND on the development of the area of *arable land*. How would the conversion process proceed in reality? How realistic is the simulation?
3. Quantify the model using consistent units (use the energy unit PJ = Peta Joule) to eliminate the presently used quaint units and their conversions (cf. e.g. Hampicke 1983, Kellner/Becker 1971, Kirchgessner 1978).
4. Apply data for (a) a developing country, and (b) the world as a whole (population increase by a surplus of births!), and investigate strategies for fighting hunger by changing the composition of the diet (vegetable vs. animal sources) and by (limited and realistic) increase of arable land.

References

Ruhrstickstoff AG 1983: *Faustzahlen für Landwirtschaft und Gartenbau.* Landwirtschaftsverlag Münster-Hiltrup, 10. Auflage.

Bundesministerium für Ernährung, Landwirtschaft und Forsten 1982: *Statistisches Jahrbuch über Ernährung, Landwirtschaft und Forsten.* Landwirtschaftsverlag Münster-Hiltrup 1982.

Hampicke, U. 1983: Die voraussichtlichen Kosten einer naturgerechten Landwirtschaft. *Landschaft und Stadt*, Vol. 15 (4), S. 171-183.

Kellner, 0., Becker, M. 1971: *Grundzüge der Fütterungslehre.* 15. Auflage, Hamburg und Berlin.

Kirchgessner, M. 1978: *Tierernährung.* 3. Auflage, Frankfurt/M.

Z314 Agriculture and farm bankruptcy

Simulation task

Where people settle, nature and landscape are characterized by agriculture. Which form of agriculture dominates in a region depends partly on natural conditions, quite decisively, however, also on economic conditions, primarily the market for products, achievable prices, and the costs of agricultural production. The forms of land cultivation reach from extensive pasturing and agroforestry to professional management of huge monocultures with intensive use of machines and chemicals. Obviously the consequences for the ecological system depend decisively on the type of cultivation – they may be considerable. Because of the complexity of the economic, ecological and social relationships the development of corresponding models is appropriate. They can provide better understanding of the system, and allow simulations to study possible dynamic developments and available options.

Within the European Union (and elsewhere) large increases in agricultural productivity cause overproduction – fewer and fewer farmers suffice to satisfy food requirements. Prices remain low while the costs to farmers are increasing, causing economic conditions that force many farmers into bankruptcy. Settlement and population patterns and social structures of entire rural regions change. With measures like price support, income subsidies, production quotas, fallowing, and other approaches administrations attempt to maintain functioning rural settlements and social structures that continue to provide basic supplies of affordable food, possibilities for employment, and a decent standard of living. But despite enormous financial cost to the taxpayer, these measures have only limited success. As in other applications, simulation models can help to better understand relevant processes and to search for appropriate solutions.

Simulation model

The simulation diagram is shown in Figure Z314a. The corresponding model equations are listed in the following.

The total *production* (of a given agricultural product, e.g. wheat) arises from the number of *farms,* their (average) *farm size* and (average) *productivity* per hectare and year. *Production* must match *grain demand* which is here assumed to correspond to the constant parameter NORMAL GRAIN DEMAND (because of approximately constant total population).

If the resulting *relative supply* signals scarcity the *relative price* will increase; for a surplus the price per unit of product will drop compared to NORMAL PRICE – unless GOVT PRICE SUPPORT attempts to stabilize (or change) it. For example, if average income of farmers is too low, the price for their products could be raised by price supports.

Depending on (grain) surplus or deficit, *sales* correspond to either *grain demand* or *production*. The *income* achieved leads after deduction of *production cost* and accounting for possible *income subsidies* (DIRECT INCOME SUBSIDY) to the corresponding *net farm income*.

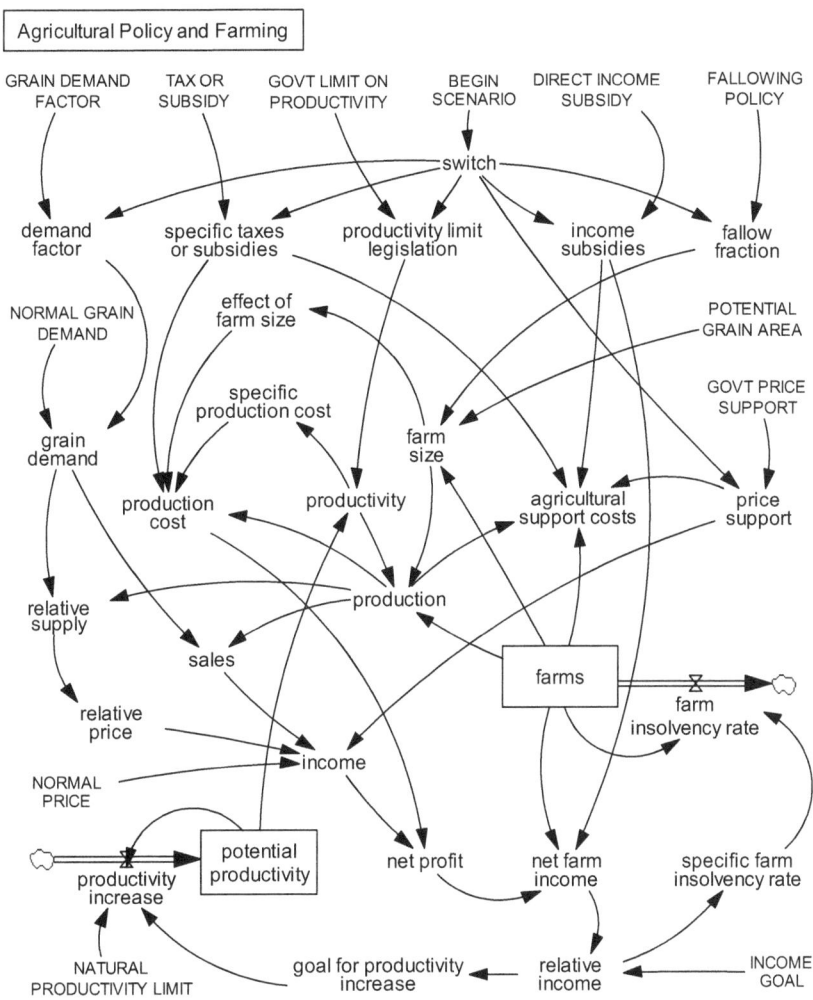

Figure Z314a: Simulation diagram for the development of farm operations.

Production cost results from (the amount of) *production* and *specific production cost*. These costs can be strongly affected by government-regulated *specific taxes* (nitrate tax, fuel tax) or *subsidies* (milk etc.) (parameter TAX OR SUBSIDY).

The farm manager orientates his decisions with respect to economic success in relation to his INCOME GOAL: If *net farm income* is lower than the INCOME GOAL, he will strive for *productivity increase*. If it is much smaller, then the probability of termination of the business, i.e. *farm insolvency rate* grows. The number of *farms* is reduced accordingly. Production can only be raised within limits provided by either NATURAL PRODUCTIVITY LIMIT or GOVT LIMIT ON PRODUCTIVITY (production quotas).

The model uses several table functions some of which strongly determine the dynamic behavior (*relative price, specific farm insolvency rate, effect of farm size, goal for productivity increase, specific production cost*). These table functions attempt to capture an assumed "average" decision behavior. This introduces an inherent uncertainty, since decisions are usually – at least partly – an expression of free will. The model user should therefore check these table functions critically and perhaps modify them according to his own well-founded conclusions.

Scenario parameters
NORMAL PRICE = 160 [$/t]
NORMAL GRAIN DEMAND = 2.5e +007 [t/Year]
BEGIN SCENARIO = 2010 [Year]
GRAIN DEMAND FACTOR = 1 [1]
TAX OR SUBSIDY = 0 [$ /t] *taxes: positive, subsidies: negative sign!*
GOVT LIMIT ON PRODUCTIVITY = 10 [t /(Year *ha)]
DIRECT INCOME SUBSIDY = 0 [$ /(farm*Year)]
FALLOWING POLICY = 0 [1]
GOVT PRICE SUPPORT = 0 [$/t]
INCOME GOAL = 30000 [$ /(farm*Year)]
NATURAL PRODUCTIVITY LIMIT = 10 [t/(Year *ha)]
POTENTIAL GRAIN AREA = 5e+006 [ha]

Dynamics
switch = STEP(1, BEGIN SCENARIO) [1]
demand factor = 1 +switch *(GRAIN DEMAND FACTOR -1) [1]
specific taxes or subsidies = switch *TAX OR SUBSIDY [$/t]
productivity limit legislation = 10 +switch *(GOVT LIMIT ON PRODUCTIVITY -10)
      [t/(Year*ha)]
income subsidies = switch *DIRECT INCOME SUBSIDY [$/(Year*farm)]
fallow fraction = switch *FALLOWING POLICY [1]
price support = switch *GOVT PRICE SUPPORT [$/t]
grain demand = NORMAL GRAIN DEMAND *demand factor [t/Year]
sales = IF THEN ELSE(grain demand<production, grain demand, production) [t /Year]
relative supply = production /grain demand [1]
relativer Preis = WITH LOOKUP (relative supply, ([(0, 0) -(5, 20)], (0.2, 15), (0.5, 5),
      (0.8, 2), (1, 1), (1.2, 0.8), (1.5, 0.5), (2, 0.3), (5, 0.3))) [1]
income = (relativer Preis *NORMAL PRICE +price support) *sales [$/Year]
net profit = income -production cost [$/Year]
net farm income = (net profit /farms) +income subsidies [$/(Year* farm)]
relative income = net farm income /INCOME GOAL [1]
specific farm insolvency rate = WITH LOOKUP (relative income, ([(0, 0) -(2, 0.5)], (0,
      0.4), (0.3, 0.15), (0.5, 0.08), (0.8, 0.02), (0.9, 0.01), (1, 0), (2, 0), (5, 0))) [1 /Year]
farm insolvency rate = -specific farm insolvency rate *farms [farm/Year]
farms = INTEG (farm insolvency rate, 100000) [farm]
farm size = (1 -fallow fraction) *POTENTIAL GRAIN AREA /farms [ha/farm]
effect of farm size= WITH LOOKUP (farm size, ([(0, 0) -(500, 2)], (0, 1.2), (25, 1.1),
      (50, 1), (100, 0.9), (1000, 0.8), (2000, 0.8))) [1]
goal for productivity increase = WITH LOOKUP (relative income, ([(0, -0.1) -(2, 0.3)],
      (0, -0.1), (0.2, -0.1), (0.4, -0.075), (0.5, 0), (0.6, 0.15), (0.7, 0.2), (0.8, 0.2), (0.9,
      0.1), (1, 0.05), (1.5, 0.02), (2, 0), (5, 0))) [1/Year]
productivity increase = goal for productivity increase *potential productivity
      *(1-(potential productivity /NATURAL PRODUCTIVITY LIMIT)) [t/(Year*Year*ha)]

potential productivity = INTEG (productivity increase, 5) [t/(ha*Year)]
productivity = IF THEN ELSE (potential productivity > productivity limit legislation,
    productivity limit legislation, potential productivity) [t/(Year*ha)]
production = farms *farm size *productivity [t/Year]
specific production cost = WITH LOOKUP (productivity, ([(0, 0) -(40, 1000)], (0, 50),
    (2.5, 60), (5, 90), (7.5, 130), (10, 180), (15, 270), (25, 350), (40, 400))) [$ /t]
production cost = (specific production cost *effect of farm size+specific taxes or
    subsidies) *production [$/Year]
agricultural support costs = income subsidies *farms +(price support -specific taxes or
    subsidies) *production [$/Year]

Simulation time parameters
INITIAL TIME = 2005 [Year]
FINAL TIME = 2025 [Year]
TIME STEP = 0.0625 [Year]

Simulation results

In assessing the following simulations with model Z314 one should keep in mind that
this simple model can only provide hints concerning possible development trends for
the different scenarios. The following statements are therefore only valid for this
model and do not represent any factual assertions concerning real developments.

Figure Z314b shows simulation results for the default parameter setting. In this
case no government interventions in agricultural production are scheduled. Because of
the relatively unfavorable *net farm income*, there is, first of all, a *high farm insolvency
rate* and accordingly a significant permanent reduction of the number of *farms*. Sec-
ondly, farm managers attempt to improve *net farm income* by *productivity increase*.
Although *net farm income* can be kept roughly constant, this is only because the number of *farms* is reduced and the (average) *farm size* increases.

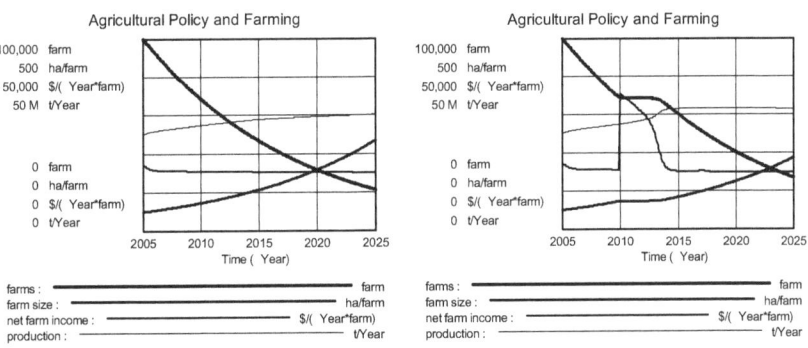

Figure Z314b: Development of farm operations without government interventions.
Figure Z314c: Development if farms receive direct income subsidies.

Figure Z314c shows results for a scenario where – beginning in year 2010 –
farms receive direct income subsidies of 20'000 $ per year. Initially this approach
succeeds in halting widespread farm bankruptcy and to secure a reasonable *net farm*

income. However, operational economic conditions deteriorate again later, since the *relative price* drops because of high *relative supply* and *net farm income* drops again to its original value despite *income subsidies*. Widespread farm insolvencies cannot be staved off by direct subsidies.

Figure Z314d shows results for a scenario where the grain price is supported with a GOVT PRICE SUPPORT of $ 80/t beginning in year 2010. This approach also can only hold back widespread farm bankruptcy for a while. After some years farm insolvencies resume, accompanied by corresponding increase in average *farm size*.

Besides these two measures which require considerable *agricultural support costs* further measures are conceivable which do not require subsidies.

Figure Z314e shows results for a scenario where part of the acreage for grain is taken out of production (FALLOWING POLICY = 20 percent = 0.2) beginning in year 2010. The corresponding reduction of production and its effect on *relative supply* lead to higher *relative price*, improving the operational economic conditions, and causing significant increase in *net farm income*. However, this approach also does not provide a lasting solution. After some years previous conditions resume again, and farm bankruptcies again reduce the number of *farms*.

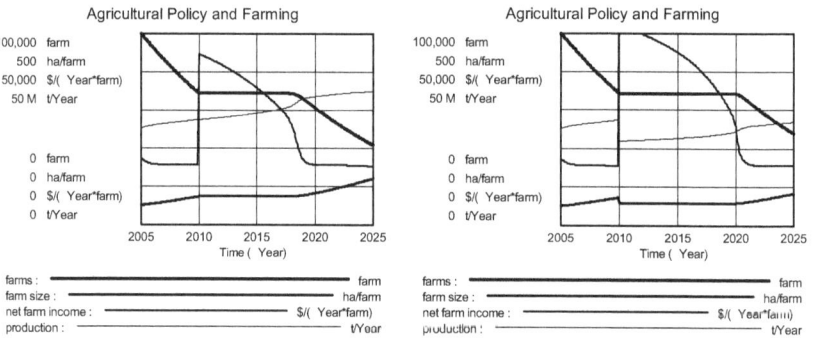

Figure Z314d: Development for grain price support.
Figure Z314e: Development if arable land is taken out of production.

Figure Z314f shows results for a scenario where the GRAIN DEMAND FACTOR was assumed to rise to 1.2 beginning in year 2010 (perhaps because of higher export demand or the use of grain for biofuel). The development also does not lead to permanent stabilization of the number of *farms* but yields a development similar to those in the other scenarios.

Figure Z314g shows results for a scenario where beginning in year 2010 further intensification of agricultural production is not allowed and productivity is restricted by GOVT LIMIT ON PRODUCTIVITY to 5 t/ha. (In practice this could be achieved by widespread extensification or introduction of organic agriculture which does not use e.g. mineral nitrogen fertilizer.) This leads to a (small) reduction of *production*, which in turn causes an increase in *relative price* and in *net farm income*. The pressure for termination of the farm business then does not exist any more. Since the *productivity increase* is also limited, the *relative price* remains high enough as a result of the *pro-*

duction being adapted to *grain demand*. This secures a favorable *net farm income* and largely prevents termination for economic reasons. As a result the average *farm size* also remains approximately constant.

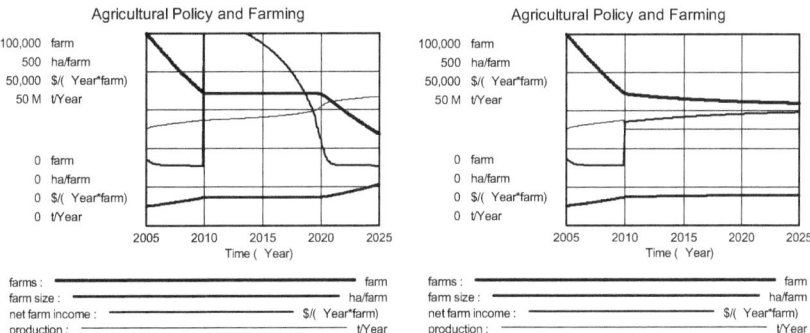

Figure Z314f: Development for increasing grain demand.
Figure Z314g: Development for limitation of hectare productivity (extensification).

Exercises

1. Examine (individually) the consequences of the different measures provided in the model, particularly as function of their strength. Follow the influence chain in the model, and determine the reasons why the measures examined in Figures Z314c to f do not lead to lasting success.

2. Investigate a "subsidies" scenario: Develop a well-coordinated agricultural policy using the parameters TAX OR SUBSIDY, DIRECT INCOME SUBSIDY, GOVT PRICE SUP- PORT. The INCOME GOAL should correspond to normal labor income. Can you find a solution which permanently guarantees farmers a good *net farm income* while main- taining rural and village structures (medium-sized farms: *farm size*)? How much does this cost the taxpayer (*agricultural support costs*)?

3. Investigate a "limited production" scenario. Develop a well-coordinated agricultural policy using the parameters GRAIN DEMAND FACTOR, GOVT LIMIT ON PRODUCTIVITY, FALLOWING POLICY, again ensuring good *net farm income* and preservation of me- dium-sized farm (*farm size*). Can you find an acceptable long-term solution preventing production surplus, price collapse, and farm bankruptcies?

4

Ecosystems and Resources

Conservation laws do not apply to organisms, populations, and resources: They can die, be extinguished or become exhausted. The conservation laws for energy and matter which formed the backbone of the simulation models in Part 3 of the System Zoo can be applied only to metabolic processes of organisms and populations and to the conversion of resources.

For the maintenance of life all organisms and populations need species-specific resources which supply them with energy and nutrients, construction material and operating supplies. Resources can be nonrenewable substances and energies like minerals or fossil fuels; humans make extensive use of such resources. Resources can also be renewable substances and energies like clean water, wood and fibers, wind and solar energy and in particular other organisms – animals and plants which are a food source for other organisms. In ecosystems innumerable "predator" populations feed on species-specific "prey" populations. The processes of "eat or be eaten" usually extend over several layers of the food-chain – always at the expense of energy losses which, however, can ensure maintenance of the populations as long as the life-supporting resource is not overused. In the course of evolution relatively stable equilibrium states have developed which stabilize complete ecosystems in the long run. If they are disturbed, e.g. by interventions of humans, then even "small" disturbances can lead to permanent collapse of whole ecosystems. Oscillations are also possible in ecosystems; they are often even part of the "natural" behavior.

In this part of the System Zoo 18 simulation models are introduced which deal with the dynamics of predator-prey relationships, food-chains in ecosystems and the use of renewable and nonrenewable resources. Models of this type – and in particular the insights which arise from the detailed representation of relationships and dependencies in complex ecosystems and the study of the dynamics of resource use and exploitation processes – have considerable significance for the protection of species and natural resources and for the sustainable use of resources to which humankind owes its existence on earth.

Z401 Predator and unlimited prey. The more predators there are, the more prey is caught, and the faster the predator population can multiply – as long as the food supply is sufficient. But predation reduces the prey population, food for the now large and growing predator population becomes increasingly rare, and its population growth reduces. Finally the predator population shrinks again; the prey population remains relatively unmolested now and can increase again. With more prey available, the predator population will again increase, and the cycle repeats – the two populations oscillate with a certain phase shift but with the same period. The model examines the dynamics for the case where the prey population can grow without limit (e.g. because of unlimited food supply). In this case undamped oscillations arise which constantly repeat with the same amplitude. Both populations cannot die out since the predators are dependent on the prey population as their only food source.

Z402 Predator and limited prey. The behavior of the predator-prey system changes drastically if the size of the prey population has an upper limit because e.g. the grazing capacity is limited in a given area. This system still produces oscillations, but they are damped and will eventually come to rest at constant equilibrium values for predator and prey. If the grazing capacity is reduced to below a certain limit, then the predator population collapses and disappears completely. Changes at the lower levels of a food-chain can therefore have dramatic consequences for organisms at higher levels although their food supply is not directly affected.

Z403 Predator and two prey populations. If two or more prey populations are available to a predator as a food source, then it appears at first logical to suspect that the system will be more stable because of the greater food variety. In fact, however, all prey populations except the most robust one are threatened with extinction because in this system the predator population does not depend on the "second to last" prey population, and its disappearance does not have any consequences for the predator. This system also oscillates. In the end, only a single predator-prey relationship remains consisting of the predator and the remaining prey population.

Z404 Prey and two predator populations. If two predator populations are dependent on the same prey population, then a reduction of prey would also mean a decline in both predator populations. The prey population is therefore preserved. The system again produces oscillations. However, in the long run the predator with a relative disadvantage disappears (e.g. the one with relatively higher food consumption). In the long run its population cannot compete against the more efficient competitor. Again only a single predator-prey relationship survives eventually. Of course there are usually more complex feeding relationships in real ecosystems – but the idealized systems considered here provide useful indications of possible dynamic developments.

Z405 Ecosystem collapse. Simple cause-effect reasoning can easily lead to completely wrong conclusions in complex systems like ecosystems. Seemingly prudent interventions can have dramatic consequences. The model simulates a historical event: To protect herds of cattle in Arizona, predators (pumas, coyotes, wolves) were systematically eliminated. The consequence was a huge increase of the previously relatively insignificant deer population which permanently destroyed the vegetation of the area (and thus also the grazing potential for cattle) before it stabilized at a small remaining population.

Z406 Birds, insects and forest. This model also simulates (simplified) an actual ecological episode (from a region in Australia). As woodlands were increasingly converted to pastures, populations of pest insects increased so strongly that the remaining woods were also quickly destroyed by them. Here again, humans ignorant of ecosystem relationships had changed a complex system and destroyed a well-adjusted ecological equilibrium. The crucial relationships were recognized only belatedly: Insects need the woodlands as a food source and grassland for maturing of their larvae. In turn birds feed on the insects, but they need the woods for their nesting sites. If the woods are increasingly destroyed, then the conditions deteriorate for the birds and improve for the insects. As predation by birds disappears, the insect population gets out of hand and destroys the remaining woods.

Z407 Plant competition and nutrient cycle. Competitive plant populations are dependent on the same nutrient stock in the soil, which they partly take up, store in their biomass and later return to the soil with leaf litter and dead biomass and its decomposition and mineralization. The efficiency of nutrient storage plays an important role in the competition of different plant species for the same nutrient stock. The plant population which stores more nutrients and produces less waste dominates in the long run. Pioneer species (short-lived, fast growing r-strategists) have an initial advantage, but in the end climax species (long-living k-strategists such as trees) will dominate.

Z408 Fish pond. In Chinese pond farming several ecological processes have been cleverly coupled for millennia to produce a maximum of food in a closed nutrient cycle. Human and animal wastes promote the growth of algae and water hyacinths in artificial ponds, fish feed on algae, and water hyacinths serve as pig fodder. Water and mud from the pond – both nutrient-rich – are used for irrigation and fertilization of fields and vegetable patches. The nutrients in food and fodder return to the system as wastes dumped into the ponds. The model computes the dynamics of these fish ponds and their fish yield over the course of several years and as function of organic and mineral fertilization. Over-fertilization will lead to eutrophication and algal blooms.

Z409 Fishery dynamics. The size of the fish population in a body of water depends via food chains on solar radiation and the nutrient supply, i.e. the supply of phytoplankton. It is therefore limited but can be sustainably harvested if overfishing and subsequent collapse of the fish population is avoided. Fishermen face the challenge of maximizing their harvest yield without causing great fluctuations or even the collapse of the fish population, and without risking heavy economic losses (e.g. by an oversized fishing fleet). If analyzed more closely, the coupled dynamics of fish population and boat stock prove to be identical with the classic predator-prey problem. As long as the catch success depends merely on fish density, the fish population is protected from being wiped out permanently. If, however, the fishermen manage to track down hidden fish swarms with modern locator technology (sonar), then this protective system structure is fundamentally changed. To protect the fish population from disappearing completely, strict catch limits must be introduced in this case.

Z410 Fishery with optimization. The proceeds from the sale of the fish catch must partly cover the expenses for procuring, maintaining and operating fishing boats and for paying boat crews. A fishery enterprise faces the problem of choosing the number of its boats and crews such that the available fish population can be used in an economically optimal and sustainable manner. If the number of fishing boats is too small, the resource can only partly be utilized; if too many fishing boats operate, the profit margin becomes too small or negative. As in other decisions of this type, the question arises what criteria are to be used for optimization: Should profit be maximized? Or should a maximum of food be obtained for a hungry population? Or shall as many permanent jobs as possible be secured? Or should different criteria with different weightings be taken into account simultaneously? The optimization criteria applied have a decisive influence on the result.

Z411 Tourism. Regions which excel with their natural beauty or cultural monuments attract tourists. But tourists need infrastructure: hotels, supplies, waste disposal, streets, and airports. The more tourists, the more infrastructure is needed, the more the region loses its original attractiveness. Advertising can partly make up for losses. Since natural and cultural attractions cannot be increased arbitrarily, there are limits to every tourist development. In the most favorable case a region manages to safeguard its advantages sustainably by careful and moderate development. Careless development will permanently destroy the base for profitable tourism.

Z412 Tourism dynamics. Systems of physics and engineering (such as vehicle suspensions, flight dynamics or electronic circuits) can (almost always) be represented in mathematical models which are provably "right" or "wrong". In "right" models processes and relationships are described by valid and proven methods and are coupled in a verifiable way. Errors are found and corrected by repeated critical checks. For model development in other disciplines such a straightforward approach is seldom possible, often for the simple reason that the complexity of the modeled system requires many simplifications which are open to argument and often not clearly decidable at all. In such cases competitive model development by working groups with thoroughly different views and experiences often provides a comprehensive view of a problem area. If despite different starting positions different system studies lead to comparable statements and perhaps even to very similar mathematical models, then these results must be taken seriously. Four different simulation models of tourism dynamics which were developed independently by four different working groups using an identical problem formulation are introduced here. Although the models are different, they lead to the same conclusions about tourism development dynamics.

Z413 Forest clearing. Agriculture and cattle grazing withdraw nutrients from the soil with the field crops and animals produced there. If originally the soil is low in nutrients – like most soils of the tropics – the soils lose their fertility sooner or later as a result of the nutrient removal. A population which has grown quickly thanks to the initially good food supply then faces food problems and must starve or leave the area. Slash and burn cultivation of fields has developed in many parts of the earth as a result of these constraints. Woods are cleared; the field is used for crop production for some years. If fertility decreases, the field is abandoned, letting natural succession take over. Eventually, original forest vegetation will take over, and in the course of decades nutrients accumulate again. After renewed clearing the area can be used again for some years for crop production.

Z414 Resource discovery. The discovery and exploitation of resources like metals or fossil fuels follows a characteristic dynamics although random events and probabilities imply significant uncertainties for individual finds. If there is demand, and large amounts of resources are still undiscovered, the search for new sources is very successful at first. The amount of discovered resources grows quickly. As more and more supplies are discovered, the amount of undiscovered resources shrinks, and the rate of discovery of new resources drops. Resource consumption increases at first with increasing resource availability, but decreases as resource stocks become exhausted.

Z415 Resource extraction and recycling. The resource mining rate is determined by demand and also by still available resources. However, the production of goods from this resource does not depend on the mining rate but on the resource supply offered in the market, which for a large part may originate from recycled material (i.e. metal scrap). The amount of recycled resource depends on the fraction of material that is recycled, and on annual scrapping rate – which is a direct function of the life time of products. The model can be used to study influences of recycling rate and of product life time on long-term resource supply, among other things. By improvement of these factors a precarious resource shortage can often be moved far into the future.

Z416 Overshoot and collapse. For renewable resources (like water, soils, food, wood, renewable energies) other conditions apply than for nonrenewable resources. Renewable resources can in principle be permanently available – but only if their use does not exceed critical limits and resources can regenerate. The dynamics is therefore primarily determined by two developments: the renewal of the resource and how it is used. The regeneration rate depends fundamentally on the remaining amount, while the utilization rate depends on the user population (e.g. grazing animals, humans) and its specific resource consumption. For different parameter combinations qualitatively different behaviors arise: equilibrium states, oscillations, limit cycle or complete destruction of the resource base and collapse of the user population.

Z417 Tragedy of the commons. If a renewable resource is common property (commons = e.g. common fishing grounds or pastures), the temptation for the individual user is to increase his personal benefit by an additional investment (e.g. a bigger boat or another head of cattle). The process can be self-amplifying if by the additional investment further profits can be achieved which allow further investments in resource-using means. The development finally leads to destruction of the common resource with corresponding losses for everyone. Sustainable use of the commons is only possible if strict rules are adhered to by all.

Z418 Sustainable use of a renewable resource. Profitable and sustainable use of a common renewable resource is possible if its use is not determined by individual profit-seeking but by a common goal of permanent preservation of the resource stock. This principle of sustainability has guided forest management in many countries for several hundred years.

Z401 Predator and unlimited prey

Simulation task

Records kept by the Hudson-Bay company of Canada over more than nine decades since 1845 on pelts received of lynxes and snowshoe hares reveal strong and regular oscillations with a period of about 9.6 years (Kormondy 1976). When becoming aware of similar periodic oscillations of fish populations in the Adriatic Sea, Vito Volterra in 1926 formulated a mathematical model which describes the dynamics of such predator-prey systems. The same mathematical formulation was developed independently in 1925 by Alfred Lotka. Their work is now known as "Lotka-Volterra systems".

If one looks separately at the populations of predator and prey, the linear model of growth and decay applies to each population (as formulated e.g. in model Z103 "Exponential growth and decay" of *System Zoo 1*). The change of population is then proportional to its respective size and relative net growth rate. Assuming unlimited grazing capacity, the prey population without predators increases exponentially with growth rate a, while the predator population without prey decreases exponentially by starvation with decline rate d. The differential equations for the rates of change of prey population x and predator population y are:

$$dx/dt = a\,x$$

$$dy/dt = -\,d\,y$$

The special dynamic properties of a predator-prey system are based on the fact that now the two populations are coupled nonlinearly with each other. Losses of the prey population by predation, and the corresponding gains of the predator population by eating the prey will depend on the frequency of the encounters between predator and prey which is proportional to the product of both populations (xy): The chances of a predator encountering prey will increase with the size of each population. In some of these encounters, prey will be caught and eaten by predators. The prey population will be reduced (factor b), while the biomass of the predator population increases accordingly (factor c). The differential equations of the Lotka-Volterra system are therefore:

$$dx/dt = a\,x - b\,xy$$

$$dy/dt = c\,xy - d\,y$$

This simple basic model can be improved by adding other terms to account e.g. for effects of limited grazing capacity, of competition among predators, of a surplus of preys, of time delays and random effects (cf. Smith 1975, Bazykin 1976, and the followings models Z402 through Z404). The nonlinear coupling between populations leads to dynamics that cannot be found in linear systems.

Volterra summarized his findings about the behavior of a predator-prey system in its simplest formulation in three "laws" (D'Ancona 1939):

Law of the periodical cycle: The fluctuations of two species populations, where one feeds on the other, are periodic, and the period depends entirely on the coefficients of growth and decay (a and d) and initial conditions (x_0, y_0).

Law of the conservation of averages: The averages of population numbers of both species remain constant and independent of the initial values of both populations if and only if the coefficients of growth and decay and the conditions of predation (prey losses, predator gains), i.e. the four coefficients *a*, *d*, *b*, *c* remain constant.

Law of perturbation of averages: If individuals of *both* species are removed (e.g. by predation) uniformly and in proportion with their total population, the average population of the prey increases, while the average population of the predator decreases. On the other hand, increased protection of the prey species will lead to growth of both populations.

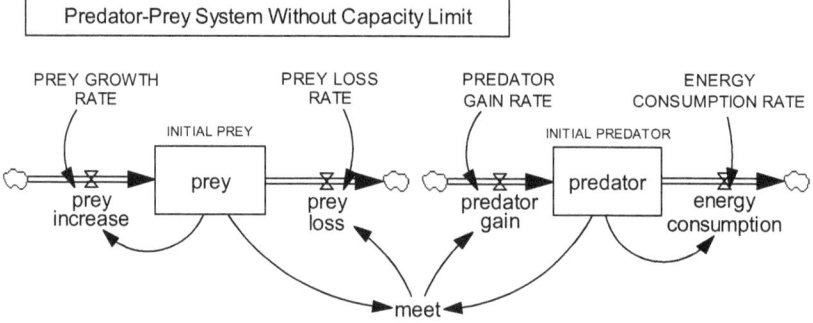

Figure Z401a: Simulation diagram of the predator-prey system with unlimited grazing.

Simulation model

The simulation diagram shown in Figure Z401a corresponds to the simple Lotka-Volterra system mentioned. The corresponding model equations are listed in the following. Parameters and state variables are normalized to unity "1". This has an advantage for the analysis of the system's dynamics, but for concrete applications all parameters, variables, and initial conditions must be replaced by real quantities (see model Z402B).

The *prey* population increases by *prey increase* which depends on (relative) PREY GROWTH RATE and *prey*. The *prey* population decreases by *prey loss* corresponding to the frequency of encounters (*meet*) between *predator* and *prey* and the PREY LOSS RATE. The *predator* population profits from predation by *predator gain* (biomass expressed as energy) contributing to maintenance and growth of the *predator* population (biomass) and compensating for its *energy consumption*. This depends on the size of the *predator* population and its (relative) ENERGY CONSUMPTION RATE.

Parameters and initial values
INITIAL PREY = 0.1 [prey biomass]
INITIAL PREDATOR = 0.1 [predator biomass]
PREY GROWTH RATE = 1 [1/Month]
PREY LOSS RATE = 1 [1/(predator biomass *Month)]
PREDATOR GAIN RATE = 1 [1/(prey biomass*Month)]
ENERGY CONSUMPTION RATE = 1 [1/Month]

Dynamics
meet = predator *prey [prey biomass* predator biomass]
prey loss = PREY LOSS RATE *meet [prey biomass /Month]
prey increase = PREY GROWTH RATE *prey [prey biomass e/Month]
prey = INTEG (+prey increase -prey loss, INITIAL PREY) [prey biomass]
energy consumption = ENERGY CONSUMPTION RATE *predator [predator biomass /Month]
predator gain = PREDATOR GAIN RATE *meet [predator biomass /Month]
predator = INTEG (+predator gain -energy consumption, INITIAL PREDATOR) [predator biomass]

Simulation time parameters
INITIAL TIME = 0 [Month]
FINAL TIME = 20 [Month]
TIME STEP = 0.02 [Month]

Simulation results

Figure Z401b shows a time plot of simulation results for the default parameter setting. If – as in this case – the system starts with low initial values for both populations, there is at first exponential growth of the prey population. As the amount of available prey increases rapidly, it allows a corresponding increase of the predator population and predation, leading to a decrease of the prey population. With some time delay, the predator population also decreases as a consequence. Undamped oscillations with a period of about 10 time units develop around an equilibrium state of the system. The prey population cannot vanish completely in this case, since the predator has no alternative source of predation and its population (and hence predation) diminishes as the prey population decreases. In this system of total dependence the prey population is inherently protected from extinction.

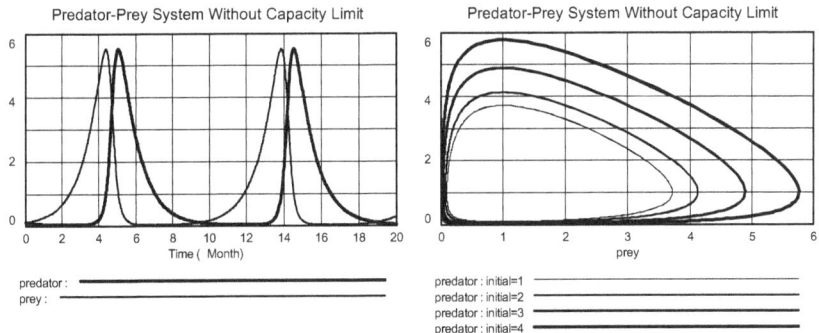

Figure Z401b: For unlimited carrying capacity the predator-prey systems oscillates without damping.
Figure Z401c: In a state diagram this yields closed cycles dependent on initial state.

Figure Z401c (state diagram) shows the development of the *predator* population as function of *prey* population for different values of INITIAL PREDATOR (1, 2, 3 and 4; other parameters as in the default setting). The development proceeds on closed, permanently recurring cycles whose "width" is determined exclusively by the initial state. The state trajectory permanently cycles around an equilibrium point of the system. The system can remain in this state (without oscillation) only if its initial state corresponds exactly to this equilibrium state (see exercises)

Exercises

1. Discuss the behavior in the time plot and also in the state diagram. Find the direction of motion in the state diagram and mark the direction with an arrow in Figure Z401c.
2. Examine the influence of the different parameters on the dynamic behavior of the system. Which parameters change the frequency of the oscillation? In which way? Which parameters determine the time delay between prey maximum and predator maximum? Explain the result!
3. Determine analytically the equilibrium points by applying the condition that the rates of change disappear (cf. Bossel *Systems and Models* 2007, 127-129). Which parameters determine the equilibrium state? Confirm the results by simulations.
4. Replace the normalized parameters of the default setting by more realistic parameters. Use plausible estimates for e.g. (a) rabbits and foxes, (b) mice and owls, (c) trout and pike. Does the fundamental behavior of the system change although parameter values differ by several orders of magnitude? Are the oscillation periods found by you realistic? (If not: review your parameter choice!)
5. Confirm the three laws of Volterra by simulations (see above).

References

Bazykin, A.D. 1976: *Structural and dynamic stability of model predator-prey systems*. IIASA, Laxenburg bei Wien, RN-76-8.
Bossel, H. 1985: *Umweltdynamik – 30 Programme für kybernetische Umwelterfahrungen*. TeWi, München, S. 91-102.
Bossel, H. 1992: *Simulation dynamischer Systeme – Grundwissen, Methoden, Programme*. Vieweg, Braunschweig und Wiesbaden, S. 63-68.
Bossel, H. SDS 2004: *Systeme, Dynamik, Simulation – Modellbildung, Analyse und Simulation komplexer Systeme*. Books on Demand, Norderstedt, S. 147-148.
Bossel, H. Zoo1 2004: *Systemzoo 1 – Elementarsysteme, Technik und Physik*. Books on Demand, Norderstedt.
D'Ancona, U. 1939: *Der Kampf ums Dasein – Eine biologisch-mathematische Darstellung der Lebensgemeinschaften und biologischen Gleichgewichte*. Bornträger, Berlin.
Kormondy, E. J. 1976: *Concepts of Ecology*. Prentice-Hall, Englewood Cliffs, N.J.
Lotka, A. J. 1956: *Elements of mathematical biology*. Dover, New York.
Richter, O. 1985: *Simulation des Verhaltens ökologischer Systeme – Mathematische Methoden und Modelle*. VCH Weinheim.
Smith, J. M. 1975: *Models in Ecology*. Cambridge University Press.
Volterra, V. 1931: *Leçon sur la théorie mathématique de la lutte pour la vie*. Gauthier-Villars, Paris.
Wissel, C. 1989: *Theoretische Ökologie*. Springer, Berlin /Heidelberg /New York.

Z402 Predator and limited prey

Simulation task

In model Z401 "Predator and unlimited prey" there is no growth restriction for the prey population – its food source is unlimited. Without predators in the system the prey population would grow exponentially. The presence of predators prevents such growth. Periodic undamped oscillations of both populations arise, where the predator population follows the prey population with some time delay (cf. results for Z401).

In reality the food supply per area is always limited – primarily on account of limited solar energy flux per area, which at intermediate latitudes amounts to about $1000 \ kWh/(m^2 \ yr)$ leading to net biomass increase (net primary productivity) of about $5 \ kWh/(m^2 \ yr)$. Any grazing must remain within these limits. At higher grazing rate, the grazed vegetation cannot regenerate and will disappear after some time, which would also lead to collapse of the population of grazing animals. In the evolution of natural ecosystems a corresponding approximate equilibrium between grazed vegetation and grazing animal population has established itself which is defined by the limited grazing capacity (carrying capacity of the ecosystem). In simulation models this limitation should be taken into account. A sensible approach is the assumption of logistic saturation (as in models Z109 and Z110 "Logistic growth" in *System Zoo 1*).

Two models of identical basic structure are examined in the following. They differ from predator-prey system Z401 by the introduction of a logistic growth function with capacity limitation of the prey population. While model Z402A otherwise corresponds to model Z401, a time function was introduced in model Z402B for change of the grazing capacity. These modifications lead to considerable qualitative changes of the behavior of the predator-prey system.

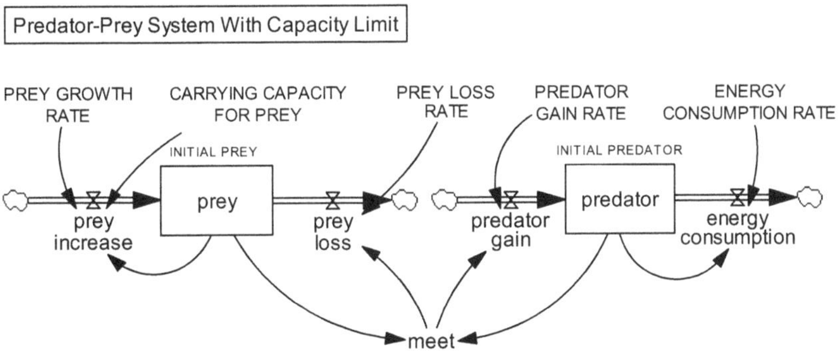

Figure Z402a: Simulation diagram for the predator-prey system with limited prey.

Simulation model Z402A

Figure Z402Aa shows the simulation diagram for this model. The respective model equations are listed in the following. Again, variables normalized to unity "1" are used. The model differs from model Z401 only by the logistic growth function for *prey increase*, which limits the prey population to the CARRYING CAPACITY FOR PREY.

Parameters and initial values
INITIAL PREY = 0.1 [prey biomass]
INITIAL PREDATOR = 0.1 [predator biomass]
CARRYING CAPACITY FOR PREY = 2 [prey biomass]
PREY GROWTH RATE = 1 [1/Month]
PREY LOSS RATE = 1 [1/(predator biomass *Month)]
PREDATOR GAIN RATE = 1 [1/(prey biomass *Month)]
ENERGY CONSUMPTION RATE = 1 [1/Month]

Dynamics
meet = predator * prey [prey biomass * predator biomass]
prey increase = PREY GROWTH RATE * prey *(1 – prey / CARRYING CAPACITY
    FOR PREY) [prey biomass /Month]
prey loss = PREY LOSS RATE * meet [prey biomass /Month]
prey = INTEG (+prey increase - prey loss, INITIAL PREY) [prey biomass]
predator gain = PREDATOR GAIN RATE * meet [predator biomass /Month]
energy consumption = ENERGY CONSUMPTION RATE * predator [predator biomass
    /Month]
predator = INTEG (+predator gain - energy consumption, INITIAL PREDATOR)
    [predator biomass]

Simulation time parameters
INITIAL TIME = 0 [Month]
FINAL TIME = 20 [Month]
TIME STEP = 0.02 [Month]

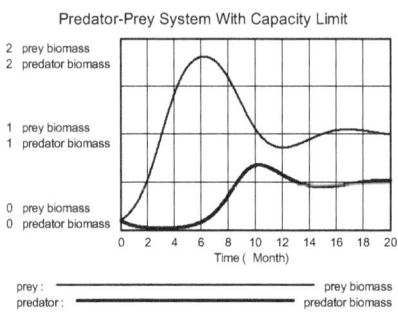 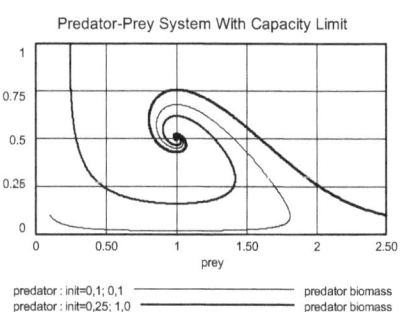

Figure Z402Ab: If carrying capacity is limited, the system exhibits damped oscillations around its equilibrium state.
Figure Z402Ac: Independently of the initial state, the state trajectories end at the same equilibrium point.

Simulation results for model Z402A

Figure Z402Ab shows the time plots for the *predator* and *prey* populations for the default parameter setting. An increase of the *prey* population is followed by growth of the *predator* population, followed by a quick reduction of the *prey* population by the large *predator* population. The process leads to periodically recurring oscillations of

both populations. Unlike the predator-prey system without capacity limit, the oscillations of the present system are strongly damped.

Figure Z402Ac shows the state trajectories for three different initial states. The
trajectory for the initial state (0.1, 0.1) (thin line) corresponds to the time plot shown
in Figure Z402Ab. In comparison to the results with model Z401 this figure also
shows that the capacity limitation causes a strongly damped motion, a spiral path ending at the equilibrium point. The state diagram shows that for quite different initial
conditions the same stable equilibrium state applies, where the populations of *predator* and *prey* can both survive indefinitely.

The system behavior changes dramatically, though, if the CARRYING CAPACITY
FOR PREY is halved (from the default value of 2 to 1). Contrary to first expectations
the *predator* population now collapses completely and permanently, while the *prey*
population grows up to its capacity limit.

In further examinations it also becomes obvious that CARRYING CAPACITY FOR
PREY determines the oscillation behavior. At low CARRYING CAPACITY FOR PREY
strongly damped oscillations arise. As the capacity parameter increases, the damping
is reduced and disappears completely if the capacity becomes infinite (this corresponds to the case without capacity limit). If the capacity drops below a certain critical
value, then the *predator* population disappears completely.

Exercises for model Z402A

1. Examine the role of the capacity parameter CARRYING CAPACITY FOR PREY in the
state diagram with *predator* as a function of *prey*. Find the critical capacity limit
where the predator population becomes extinct after some time. How does this value
depend on other parameters?
2. Determine the period of oscillation as a function of CARRYING CAPACITY FOR PREY.
3. Determine the equilibrium points analytically as function of the model parameters
(of the total of three equilibrium points, only one is stable!).
4. Couple module Z115 "State diagram" (in *System Zoo 1*) to model Z402A, generate
the state diagram (in the range corresponding to Figure Z402Ac), and confirm the
analytical result concerning the location of the equilibrium points.

Simulation model Z402B

In this simulation model the basic structure of model Z402A is translated into a more
realistic model and applied to a fox population which feeds on a rabbit population.
The simulation diagram is shown in Figure Z402Ba. The corresponding model equations are listed in the following.

The predator population *foxes* is measured in fox biomass units, the prey population *rabbits* in rabbit biomass units. Since biomass is equivalent to energy, the use of
an energy unit would be more appropriate. However, for a more vivid description the
numbers of *foxes* and *rabbits* will be used.

The biomass of *foxes* increases according to *fox gain*. It is assumed that foxes
feed exclusively on rabbits. The *fox gain* is bigger if more *encounters* take place between *rabbits* and *foxes*, and if the FOX BIOMASS GAIN PER ENCOUNTER is higher (i.e.
the bigger the hunting success per encounter).

The energy loss for maintenance of life processes of *foxes* (*fox loss*) is a function of the number of *foxes* and their (specific) ENERGY CONSUMPTION RATE OF FOXES. This maintenance demand is assumed to be 0.2/week, i.e. without food supply a fox would lose 20 percent of its weight (more correctly: its energy content) per week. In a dynamic equilibrium, the food uptake should obviously be equal to this value.

We assume here that for unhindered growth the rabbit population would have a RABBIT GROWTH RATE of 0.08/week, which amounts to doubling of the population in approximately 9 weeks. This *rabbit increase* is reduced to zero when the GRAZING CAPACITY limit is reached (e.g. 1000 rabbits). The population of *rabbits* is reduced by losses to the foxes (*rabbit loss*).

Figure Z402Ba: Simulation diagram for a system of rabbits and foxes.

The coupling between the two populations arises from the probability that the two populations meet, i.e. foxes successfully tracking down a rabbit. The more *rabbits* there are, the more are available as prey for *foxes*. The more *foxes* there are, the more *rabbits* will fall prey to foxes. The product of *rabbits* and *foxes* (= *encounter*) can therefore be taken as a measure for the number of rabbits becoming the victim of foxes. Any rabbit caught means a corresponding *fox gain* for *foxes* and a *rabbit loss* for *rabbits*. The following considerations apply to the quantification of these gains or losses: For an average number of 500 rabbits and 10 foxes (i.e. *encounter* = 5000/week) the foxes are assumed to be able to restore their maintenance energy loss (*fox loss*) of $10 \cdot 0.2 = 2$ (fox biomass per week). From this follows the (relative) FOX BIOMASS GAIN PER ENCOUNTER as $2/5000 = 0.0004$. Assuming that one fox has the biomass of 5 rabbits, the fox gain of 2 fox units per week corresponds to a rabbit loss of 10 rabbits per week. The (relative) *rabbit biomass loss per encounter* will therefore be $10/5000 = 0.002$.

To be able to examine the effect of a temporary reduction of the *max grazing capacity* (e.g. by a drought period), the time function GRAZING RESTRICTION is provided as a table function. In the default setting, the *max grazing capacity* is reduced to one half after 100 weeks, and returns to the original value after 200 weeks.

Parameters and initial values
INITIAL RABBITS = 500 [rabbit biomass]
INITIAL FOXES = 10 [fox biomass]
RABBIT GROWTH RATE = 0.08 [1/Week]
ENERGY CONSUMPTION RATE OF FOXES = 0.2 [1/Week]
RABBIT BIOMASS LOSS PER ENCOUNTER = 0.002 [rabbit biomass /(rabbit
    biomass * fox biomass)]
FOX BIOMASS GAIN PER ENCOUNTER = 0.0004 [fox biomass /(fox biomass
    *rabbit biomass)]
GRAZING CAPACITY = 1000 [rabbit biomass]
GRAZING RESTRICTION = WITH LOOKUP (Time / time unit, ([(0, 0) -(1000, 2)],
    (0, 1), (99, 1), (100, 0.5), (199, 0.5), (200, 1), (400, 1), (1000, 1))) [1]
time unit = 1 [Week]

Dynamics
max grazing capacity = GRAZING CAPACITY * GRAZING RESTRICTION
    [rabbit biomass]
free grazing capacity = 1 − rabbits / max grazing capacity [1]
encounter = rabbits * foxes / time unit [rabbit biomass * fox biomass /Week]
rabbit loss = RABBIT BIOMASS LOSS PER ENCOUNTER * encounter
    [rabbit biomass/Week]
rabbit increase = free grazing capacity * RABBIT GROWTH RATE * rabbits
    [rabbit biomass/Week]
rabbits = INTEG (+rabbit increase - rabbit loss, INITIAL RABBITS) [rabbit biomass]
fox loss = (ENERGY CONSUMPTION RATE OF FOXES) * foxes [fox biomass /Week]
fox gain = FOX BIOMASS GAIN PER ENCOUNTER * encounter [fox biomass /Week]
foxes = INTEG (+fox gain - fox loss, INITIAL FOXES) [fox biomass]

Simulation time parameters
INITIAL TIME = 0 [Week]
FINAL TIME = 400 [Week]
TIME STEP = 0.25 [Week]

Simulation results for model Z402B

Figure Z402Bb shows the temporal development of the populations of *rabbits* and
foxes for the default parameter setting over a period of 400 weeks. Figure Z402Bc
shows the state trajectory for this simulation (*foxes* as function of *rabbits*). Initially the
system moves in damped oscillation towards the equilibrium point (500 *rabbits*, 20
foxes). Before reaching this point the GRAZING RESTRICTION reduces *max grazing
capacity* to one half in week 100. The population of *rabbits* decreases initially, but
soon recovers, while the population of *foxes* disappears almost completely. Shortly
before the extinction of *foxes*, the GRAZING RESTRICTION is lifted and *max grazing
capacity* returns to its original value. This causes rapid growth of the population of
rabbits, which means an increase in the food supply for *foxes*, and a corresponding
recovery and rapid growth of their population. The foxes were lucky this time, and the
system performs its damped oscillation, coming to rest at the (original) point of equilibrium.

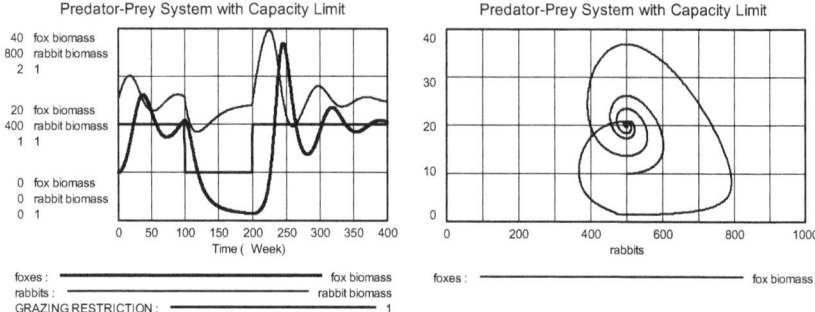

Figure Z402Bb: If grazing capacity is halved, the predator population collapses.
Figure Z402Bc: The state trajectory heads for its equilibrium in a damped oscillation.

The system oscillates at a frequency which depends on the choice of system parameters. If the grazing capacity is strongly increased (e.g. to 100'000), we observe
very strong oscillations with approximately constant amplitude, comparable to the
result for unlimited grazing capacity in model Z401. If grazing capacity is halved in
this case (applying GRAZING RESTRICTION) the population of *foxes* no longer collapses.

Exercises for model Z402B

1. Set up the simulation program and reproduce the results for the standard case with
initial conditions 500 *rabbits*, 10 *foxes*. Interpret the result and particularly the state
diagram (*foxes* vs. *rabbits*). Indicate the direction of motion of different sections with
arrows.
2. By repeated simulations with changed parameters, investigate which parameters
substantially influence the oscillation frequency of the system.
3. Examine the effect of GRAZING RESTRICTION by changing the intensity of the restriction by modifying the table function. What consequences does this have for the
predator-prey system? Does it produce qualitatively different behavior for other initial
values? What is the minimum remaining *max grazing capacity* to avoid collapse of the
predator population?
4. Introduce a seasonal (cyclical) GRAZING RESTRICTION (modify table function accordingly). How does this affect the results?
5. Increase *fox loss* temporarily (by shooting, or rabies) (keeping grazing capacity
constant). What happens? (The model requires some additions for this).
6. Use simulations to find – for constant GRAZING RESTRICTION = 1 and different initial states, primarily in the neighborhood of a suspected equilibrium point – the (three)
equilibrium points of the system. Determine the type and stability of each equilibrium
point from the course of state trajectories in its neighborhood. Sketch the state trajectories with directional arrows, showing in particular their course near equilibrium
points.
7. Obtain the system equations from the model documentation and write these (after
renaming) in mathematical notation as two differential equations. Discuss this system

of nonlinear differential equations. Determine the equilibrium points analytically and compare the result with the state diagram of the simulation.

8. Investigate how the stable equilibrium point moves if ENERGY CONSUMPTION RATE OF FOXES is changed (increased, decreased). Can you find a plausible explanation for the result?

9. Determine the linearized state equations at the three equilibrium points, introduce the corresponding system parameters into the linear system in model Z114 "Linear oscillator" (*System Zoo 1*) and check whether the behavior of the linearized system corresponds to that of the nonlinear system at the corresponding equilibrium point. (Linearization is explained e.g. in Bossel 2007: *Systems and Models*, 144-147).

References

cf. model Z401 "Predator and unlimited prey"

Z403 Predator and two prey populations

Simulation task

In models Z401 and Z402 the predator population obtained its food energy entirely from a single prey population. It turned out that this dependence protects both populations from extinction if sufficient grazing capacity is available for the prey population. The predator dies out if grazing capacity is too low, while the prey population adapts to the available grazing capacity.

However, what is to be expected if several sources of prey are at a predator's disposal, or if conversely several predator populations rely on a single prey population? This question cannot be answered reliably by intuition. Two models shall therefore be developed to investigate the dynamics of the dependence of one predator on two prey populations (Z403), and the dependence of two predator populations on a single prey population (Z404). One should keep in mind, however, that these simple models can only provide hints of behavioral trends in real ecosystems. These usually contain more complex relationships than are modeled here.

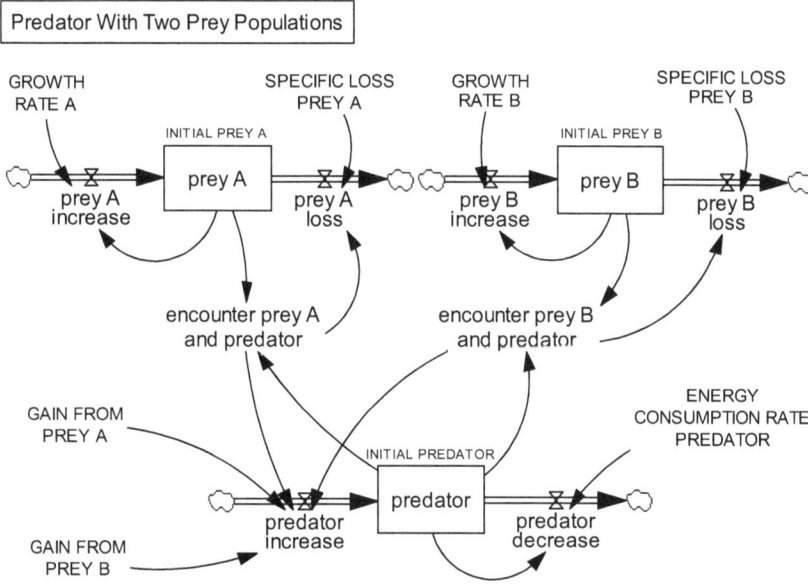

Figure Z403a: Simulation diagram for predator with two prey populations.

Simulation model

Figure Z403a shows the simulation diagram for a predator-prey system with two different prey populations. The model equations are listed in the following.

The *predator* population is coupled to *prey A* by *encounter prey A and predator* and to *prey B* by *encounter prey B and predator*. The basic structure of each of these relationships corresponds to that of the simple predator-prey system. The populations of *prey A* and *prey B* increase according to their specific GROWTH RATE A or GROWTH RATE B. The *predator* has a specific ENERGY CONSUMPTION RATE PREDATOR. The relative losses of *prey A* and *prey B* by the predation process are determined from corresponding SPECIFIC LOSS PREY A and SPECIFIC LOSS PREY B. The relative gains of the *predator* population by predation of *prey A* or *prey B* are specified by GAIN FROM PREY A and GAIN FROM PREY B.

Parameters and initial values
INITIAL PREY A = 1 [biomass A]
INITIAL PREY B = 1 [biomass B]
INITIAL PREDATOR = 1 [biomass predator]
GROWTH RATE A = 0.1 [1/Month]
GROWTH RATE B = 0.12 [1/Month]
SPECIFIC LOSS PREY A = 0.1 [(biomass A /Month) /(biomass A *biomass predator)]
SPECIFIC LOSS PREY B = 0.1 [(biomass B /Month) /(biomass B *biomass predator)]
ENERGY CONSUMPTION RATE PREDATOR = 0.1 [1/Month]
GAIN FROM PREY A = 0.1 [(biomass predator /Month) /(biomass A
        *biomass predator)]
GAIN FROM PREY B = 0.1 [(biomass predator /Month) /biomass B
        *biomass predator)]

Dynamics
encounter prey A and predator = prey A *predator [biomass A *biomass predator]
encounter prey B and predator = prey B *predator [biomass B *biomass predator]
prey A increase = GROWTH RATE A *prey A [biomass A /Month]
prey A loss = SPECIFIC LOSS PREY A *encounter prey A and predator
        [biomass A /Month]
prey A = INTEG (+prey A increase -prey A loss, INITIAL PREY A) [biomass A]
prey B increase = GROWTH RATE B *prey B [biomass B/Month]
prey B loss = SPECIFIC LOSS PREY B *encounter prey B and predator
        [biomass B /Month]
prey B = INTEG (+prey B increase -prey B loss, INITIAL PREY B) [biomass B]
predator increase = GAIN FROM PREY A *encounter prey A and predator +GAIN
        FROM PREY B *encounter prey B and predator [biomass predator /Month]
predator decrease = ENERGY CONSUMPTION RATE PREDATOR *predator
        [biomass predator /Month]
predator = INTEG (+predator increase -predator decrease, INITIAL PREDATOR)
        [biomass predator]

Simulation time parameters
INITIAL TIME = 0 [Month]
FINAL TIME = 200 [Month]
TIME STEP = 0.05 [Month]

Simulation results

Figure Z403b shows the time plot of system development for the default parameter setting. The state diagram for this development is presented in Figure Z403c. The conditions for the two prey populations differ only in a single aspect: GROWTH RATE B is insignificantly larger than GROWTH RATE A.

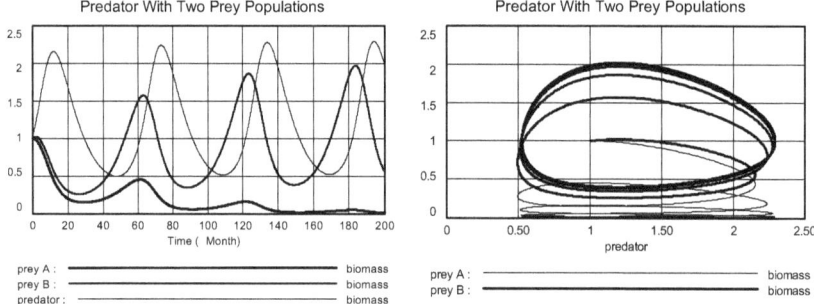

Figure Z403b: The prey population with a minor disadvantage (here *prey A*) dies out.
Figure Z403c: In the long run a predator-prey cycle develops with remaining *prey B*.

The oscillation typical of predator-prey systems is found again also for this system. However, it turns out that the prey population characterized by relative disadvantage (having e.g. lower growth rate or relatively larger predation loss rate; here *prey A*) suffers from predation at a higher rate than the more advantaged prey population. Since the predator is not dependent on the "second to last" prey population and is not affected by its disappearance, there is no protection for this prey population. Therefore *prey A* goes extinct after some time. This is generally true even if further prey populations were involved. Except for the last population all other prey populations disappear. Only the better adapted prey population (in comparison with other prey populations) can survive in the long run.

Exercises

1. Examine the development of the system for different parameter constellations. Which options does *prey A* have to survive *prey B* in the long run despite the disadvantage with respect to GROWTH RATE A?
2. Investigate how development (particularly oscillation behavior and relative survival) depends on initial values of the three populations.

References

cf. model Z401 "Predator and unlimited prey"

Z404 Prey and two predator populations

Simulation task

Often a single prey population is the source of food for several competing predators (e.g. mice as prey of foxes and birds of prey). Here again a reliable intuitive assessment of long-term development resulting from the particular system relationships is impossible. A simulation model can assist in recognizing development trends inherent in the system structure even if in reality a variety of other factors determine the development and may cause it to proceed on a somewhat different path.

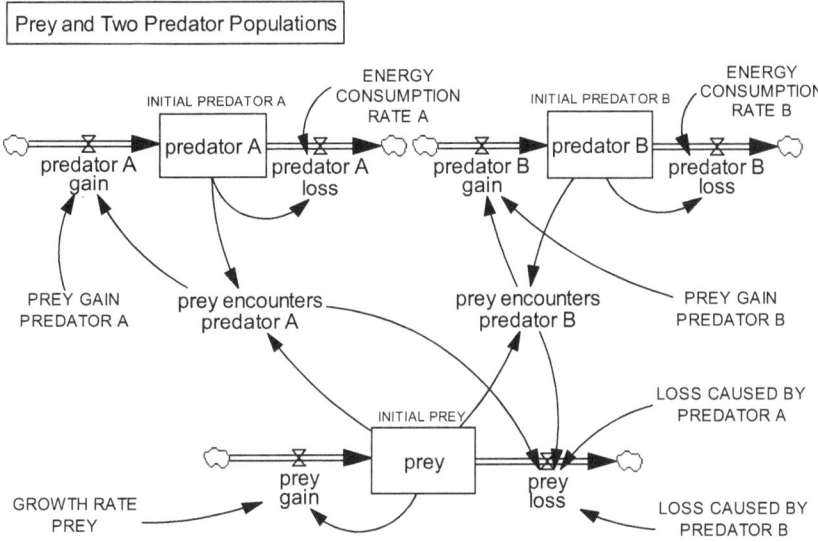

Figure Z404a: Simulation diagram for one prey and two predator populations.

Simulation model

Figure Z404a shows the simulation diagram for a predator-prey system with a single prey population and two predator populations. The corresponding model equations are listed in the following.

In this system with a single *prey* population and two predator populations *predator A* and *predator B* there exist again two predator-prey linkages (*prey encounters predator A*, and *prey encounters predator B*). These linkages correspond to those in the normal predator-prey system. *Predator A* and *predator B* have specific ENERGY CONSUMPTION RATE A and ENERGY CONSUMPTION RATE B. The prey population has (relative) GROWTH RATE PREY. The rates of increase of the *predator* populations (*predator A gain*, and *predator B gain*) are determined by (relative) PREY GAIN PREDATOR A and PREY GAIN PREDATOR B. The *prey loss* is a function of (relative) LOSS CAUSED BY PREDATOR A and LOSS CAUSED BY PREDATOR B.

Parameters and initial values
INITIAL PREY = 1 [biomass prey]
INITIAL PREDATOR A = 1 [biomass A]
INITIAL PREDATOR B = 1 [biomass B]
GROWTH RATE PREY = 0.1 [1/Month]
LOSS CAUSED BY PREDATOR A = 0.1 [(biomass prey /Month)/(biomass A
      *biomass prey)]
LOSS CAUSED BY PREDATOR B = 0.1 [(biomass prey /Month)/(biomass prey
      *biomass B)]
PREY GAIN PREDATOR A = 0.1 [(biomass A /Month)/(biomass A *biomass prey)]
PREY GAIN PREDATOR B = 0.1 [(biomass B /Month)/(biomass B *biomass prey)]
ENERGY CONSUMPTION RATE A = 0.12 [1/Month]
ENERGY CONSUMPTION RATE B = 0.1 [1/Month]

Dynamics
prey encounters predator A = predator A *prey [biomass A *biomass prey]
prey encounters predator B = predator B *prey [biomass B *biomass prey]
prey gain = GROWTH RATE PREY *prey [biomass prey /Month]
prey loss = LOSS CAUSED BY PREDATOR A *prey encounters predator A +LOSS
      CAUSED BY PREDATOR B *prey encounters predator B [biomass prey /Month]
prey = INTEG (+prey gain -prey loss, INITIAL PREY) [biomass prey]
predator A gain = PREY GAIN PREDATOR A *prey encounters predator A
      [biomass A /Month]
predator A loss = ENERGY CONSUMPTION RATE A *predator A [biomass A /Month]
predator A = INTEG (+predator A gain -predator A loss, INITIAL PREDATOR A)
      [biomass A]
predator B gain = PREY GAIN PREDATOR B *prey encounters predator B
      [biomass B /Month]
predator B loss = ENERGY CONSUMPTION RATE B *predator B [biomass B /Month]
predator B = INTEG (+predator B gain -predator B loss, INITIAL PREDATOR B)
      [biomass B]

Simulation time parameters
INITIAL TIME = 0 [Month]
FINAL TIME = 200 [Month]
TIME STEP = 0.05 [Month]

Simulation results

Figure Z404b shows the time plot of simulation results for the default parameter setting. The development of the state is represented in the state diagram of Figure Z404c.
As in model Z403 the two predator populations differ only slightly with respect to a
single parameter: The ENERGY CONSUMPTION RATE A for *predator A* is insignificantly
higher than that for *predator B*

As expected, predator-prey oscillations develop for the three populations. Since
both predator populations are dependent on a common prey, the prey population cannot possibly go extinct, since the predator populations would diminish for lack of food
if predation were too high, and the prey population is therefore ultimately protected
from extinction.

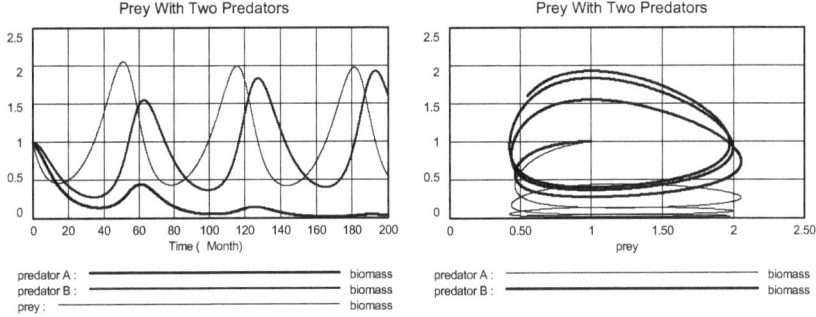

Figure Z404b: The slightly disadvantaged predator A dies out.
Figure Z404c: In the long run the state diagram shows an undamped cycle of *prey* and *predator B*.

However, a small disadvantage of *predator A* (insignificantly higher ENERGY CONSUMPTION RATE A) suffices to cause this population to disappear after some time. Thereafter an undamped oscillation develops for *prey* and the remaining *predator B* (undamped because of unrestricted carrying capacity). In the long run, the behavior of the system corresponds to that of the predator-prey system in model Z401.

Exercises

1. Examine the development of the system for different parameter constellations. What options does *predator A* have to finally survive *predator B* despite the disadvantage of ENERGY CONSUMPTION RATE A?
2. Determine how system development (particularly oscillatory behavior and relative survival) depend on initial values of the three populations.
3. Introduce a grazing capacity limit (as in model Z402). How does this affect the dynamics of the system with two predators?

References

cf. model Z401 "Predator and unlimited prey"

Z405 Ecosystem collapse

Simulation task

The disruption of a well-adjusted predator-prey relationship can have dramatic consequences. A historical example of this is the collapse of the deer population on the Kaibab Plateau in Arizona at the beginning of the 20th century (Kormondy 1976).

The Kaibab Plateau consists of about 727'000 acres (1 acre = 0.4 hectares) on the north side of the Grand Canyon. Prior to 1907 the deer herd numbered about 4000. In 1907 a bounty was placed on cougars, wolves, and coyotes, which are natural predators of deer. Within 15 or 20 years the predator population was substantially reduced; over 8000 were shot. As a consequence, the deer population erupted and had increased more than tenfold by 1918. This resulted in increasing exploitation of normal food sources. The deer turned to young timber growth, exhausting the food supply faster than it could be replaced. Although competent investigators warned of the consequences of the evident overbrowsing, no action was taken, and unimpaired by natural predation the deer population reached an estimated 100'000 in 1924. For lack of sufficient food, 60 percent of the herd died off in two successive winters. The vegetation was damaged by overbrowsing to such an extent that the food reserve could not recover and could only support half the original deer population.

With the model described here Goodman 1974 attempted to capture the dynamics of this historical event with a simulation model. The results agree quite well with the historical data.

Simulation model

The simulation diagram (Figure Z405a) shows relevant system relations. The corresponding model equations are listed in the following.

The *deer* live in a limited AREA on a *food supply* whose *food increase* is limited by its *regeneration time*. The *net increase* of *deer* depends on the amount of *food per deer* and on *deer* population. The size of the *deer* population and DAILY REQUIREMENT per deer determine *food demand*, leading to *browsing loss* of *food supply*. The *food supply* regenerates corresponding to MAX FOOD CAPACITY, but the rate of *food increase* is reduced as the *vegetation density* is reduced by overbrowsing. This is taken into account by adjusting *regeneration time* as a function of *vegetation density*.

The *deer* population is reduced by *predation loss* caused by predators (cougar, coyote, wolf) whose number is provided by the PREDATOR time function. It is assumed that as a result of the bounty the predator population is reduced linearly with time (cf. table function PREDATOR). The *predation loss* of *deer* depends on *deer density* and the number of the predators.

The model quantification is based on the following assumptions: The AREA has a size of 800'000 acres. The DAILY REQUIREMENT per deer is 2000 kilocalories per deer and day. The MAX FOOD CAPACITY of the area is 480 million [(kcal/day)/year]. Other important quantifications are incorporated in the three table functions. E.g. the function for the (net) *growth rate* of the *deer* population assumes a population decrease of 15 percent per year if the daily *food per deer* drops to 500 [kcal/day], and an annual increase of 15 percent if the daily *food per deer* reaches 1500 [kcal/day]. The table function for *predation rate* (prey per predator per year) indicates that at high *deer density* each predator kills at most 56 deer per year and correspondingly fewer deer at lower *deer density*. The table function for *regeneration time* of the *food supply* assumes that the regeneration time of normally one year can increase to 35 years if the ecosystem has been destroyed by overbrowsing.

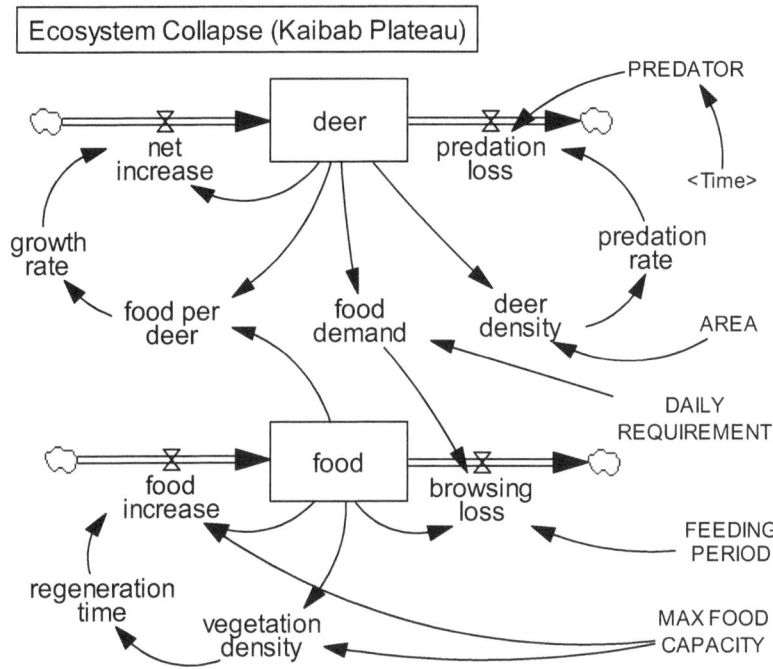

Figure Z405a: Simulation diagram for the Kaibab ecosystem.

Parameters and initial values
AREA = 800000 [acre]
MAX FOOD CAPACITY = 4.8e+008 [(kcal/Day)*Year]
DAILY REQUIREMENT = 2000 [kcal/(deer*Day)]
FEEDING PERIOD = 1 [Year]
PREDATOR = WITH LOOKUP (Time, ([(0, 0) -(50, 300)], (0, 265), (5, 245), (10, 200),
    (15, 65), (20, 8), (25, 0), (30, 0), (35, 0), (40, 0), (50, 0))) [predator]

Dynamics
deer density = deer /AREA [deer/acre]
predation rate = WITH LOOKUP (deer density, ([(0, 0) -(0.06, 60)], (0, 0), (0.005, 3),
    (0.01, 13), (0.015, 28), (0.02, 51), (0.025, 56), (0.05, 56))) [deer/(predator*Year)]
predation loss = predation rate *PREDATOR [deer/Year]
food per dear = (food /deer) [Year*kcal/(Day*deer)]
growth rate = WITH LOOKUP (food supply, ([(0, -1) -(10000, 1)], (0, -0.5), (500, -0.15),
    (1000, 0), (1500, 0.15), (2000, 0.2), (10000, 0.2))) [1/Year]
net increase = deer *growth rate [deer/Year]

deer = INTEG (+net increase -predation loss, 4000) [deer]
food demand = DAILY REQUIREMENT *deer [kcal/Day]
browsing loss = IF THEN ELSE(food demand >= (food /FEEDING PERIOD),
    (food /FEEDING PERIOD), food demand) [kcal/Day]
vegetation density = food /MAX FOOD CAPACITY [1]
regeneration time = WITH LOOKUP (vegetation density, ([(0, 0) -(1, 40)], (0, 35),
    (0.25, 15), (0.5, 5), (0.75, 1.5), (1, 1))) [Year]
food increase = (MAX FOOD CAPACITY -food) /regeneration time [kcal/Day]
food supply = INTEG (+food increase -browsing loss, 4.7e+008) [Year*kcal/Day]

Simulation time parameters
INITIAL TIME = 0 [Year]
FINAL TIME = 50 [Year]
TIME STEP = 0.25 [Year]

Simulation results

Figure Z405b shows the time plots for *food supply* and *deer* computed with the default
parameter setting, as well as the (given) time function for the PREDATOR population,
which was almost completely eradicated in the first twenty years. Figure Z405c shows
the state trajectory for *deer* as a function of *food supply*.

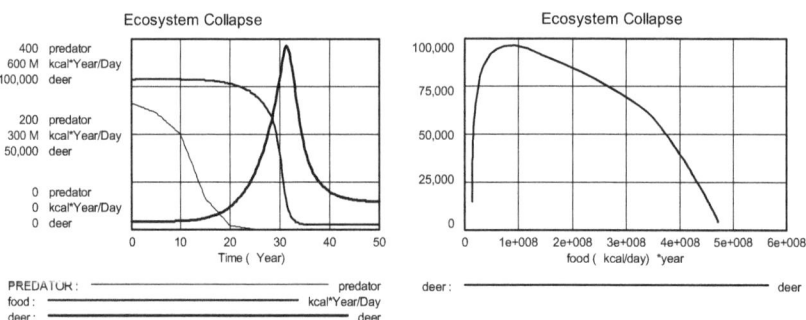

Figure Z405b: After elimination of predators the deer population increases rapidly and
then starves because of overbrowsing and destruction of its food base.
Figure Z405c: The state trajectory shows the transition from one equilibrium state
(with predators) to another one (with destroyed food base).

The computed development corresponds to historical observations: With the disappearance of predators the *deer* population can increase almost exponentially and
thereby destroy its source of *food supply* almost completely. As a consequence the
deer population collapses rapidly because of inadequate *food per deer*, and then stabilizes at a low level corresponding to the *food supply* which the degraded ecosystem
can still deliver.

Exercises

1. Was the ecosystem initially in equilibrium? What equilibrium state is reached if predators are not hunted and killed?
2. What predator control strategy should have been applied to maintain a large and stable deer population without risking collapse of the vegetation base?
3. Is it possible to return the ecosystem to its original state? How? Which irreversible changes not now represented in the model would have to be taken into account in a more complete analysis?
4. What could be achieved by again introducing predators or by managing the deer population through controlled culling?
5. Introduce a predator population as further (endogenous) state variable whose growth is determined by the supply of prey and by a hunting policy for predators provided as scenario function.

References

Goodman, M. R. 1974: *Study notes in system dynamics.* Wright-Allen-Press, Cambridge, Mass., S. 377-388.
Kormondy, E. J. 1976: *Concepts of ecology.* Prentice-Hall, Englewood Cliffs, N.J., S. 111-112.

Z406 Birds, insects, and forest

Simulation task

The state of ecosystems it determined by complex relationships between its components which have evolved in the course of evolutionary development. These are usually multiple dependence relations between organisms via nutrient cycles, food chains and food nets, predator-prey systems, symbioses, pollination, seed distribution and many other processes. The interacting dynamic processes control and regulate each other such that a dynamic equilibrium typical of the respective ecosystem evolves. Interventions which damage or enhance particular components can therefore tip the system into another state.

Such a process was observed e.g. in Australia and formulated by Trenbath and Smith 1981 as a simulation model (Richter 1985). To create larger grazing areas for sheep, eucalyptus forests in New South Wales had been cleared. Although care was taken to preserve about 20 percent of the woodlands to prevent desertification of the grassland, the remaining forest stands nevertheless collapsed in short time by insect pest attacks.

Two reasons were suspected for the disastrous multiplication of pest insects and the collapse of the remaining forests: First, (relative) enlargement of the grazing area improved conditions for the insect larvae living on grass roots. Second, reduction of forest area also meant a reduction of nesting sites for birds, reducing the population of these predators feeding on the insects.

The model therefore describes the following relationships: A region with a maximum biomass capacity K consists partly of forest x, partly of grassland vegetation (K $-x$). Birds live on insects and need the forest for nesting sites. Insects need the forest as a food source, and grassland for the maturing of larvae. If the woods are increasingly destroyed, then conditions deteriorate for the birds and improve for the insects. At a certain stage the insects get out of hand and destroy the remaining woods.

Simulation model

Figure Z406a shows the simulation diagram of the model. The model equations are listed in the following. The model consists of three modules which represent the development of the state quantities *forest biomass*, *insect biomass*, and *bird biomass*.

Forest: The structure of this module is that of a logistically growing state variable *forest biomass* with rates of change *increase forest, deforestation,* and *insect predation*. The *increase forest* is determined by GROWTH RATE FOREST and MAX BIOMASS FOREST. *Deforestation* can be adjusted by scenarios for DEFORESTATION RATE and FINAL YEAR DEFORESTATION. *Insect predation* depends on *insect biomass*, MAX PREDATION RATE INSECTS and current *forest biomass*. For the latter effect a Michaelis-Menten saturation process with HALF SATURATION CONSTANT PREDATION BY INSECTS is used (see model Z111 in *System Zoo 1*).

Insects: The basic structure is also that of logistic growth of *insect biomass*. The growth capacity is variable in this process and depends on available *forest biomass* as a measure for forest area (x) and on grassland area (K $- x$). The *increase insects* depends on *insect biomass,* the REPRODUCTION RATE INSECTS, and the CAPACITY FACTOR INSECTS. The *insect biomass* has losses by *bird predation*. These losses depend on *bird biomass*, MAX PREDATION RATE BIRDS, and – via HALF SATURATION CONSTANT PREDATION BY BIRDS – also on *insect biomass*.

Birds: The rate of *increase birds* of *bird biomass* is also determined by logistic growth with variable capacity. This depends on *forest biomass* and *insect biomass*. The CAPACITY FACTOR BIRDS and REPRODUCTION RATE BIRDS also affect the *increase birds*.

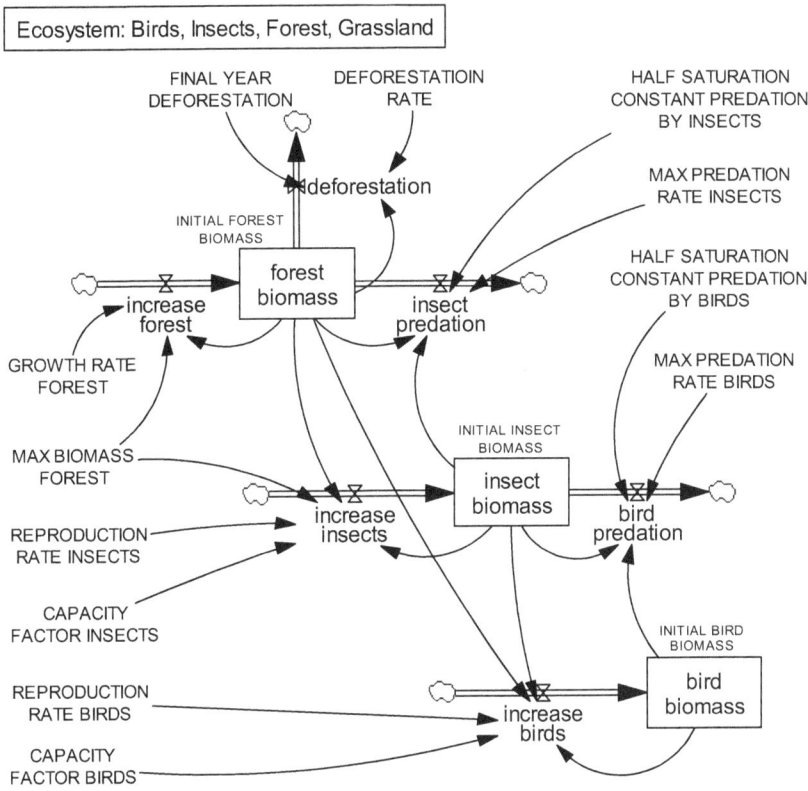

Figure Z406a: Simulation diagram for birds, insects, and forest.

Parameters and initial values
MAX BIOMASS FOREST = 100 [t ODM/ha]
GROWTH RATE FOREST = 0.1 [1/Year]
DEFORESTATIOIN RATE = 0.05 [1/Year]
FINAL YEAR DEFORESTATION = 10 [Year]
INITIAL FOREST BIOMASS = 20 [t ODM /ha]
INITIAL INSECT BIOMASS = 0.0001 [t ODM /ha]
INITIAL BIRD BIOMASS = 0.0001 [t ODM /ha]
REPRODUCTION RATE INSECTS = 2 [1/Year]
REPRODUCTION RATE BIRDS = 1 [1/Year]

CAPACITY FACTOR INSECTS = 0.0001 [1/(t ODM /ha)]
CAPACITY FACTOR BIRDS = 0.008 [ha*1/t OTS]
HALF SATURATION CONSTANT PREDATION BY INSECTS = 1 [t ODM /ha]
HALF SATURATION CONSTANT PREDATION BY BIRDS = 0.001 [t ODM /ha]
MAX PREDATION RATE INSECTS = 365 [1/Year]
MAX PREDATION RATE BIRDS = 30 [1/Year]

Dynamics
increase forest = GROWTH RATE FOREST * forest biomass *(1 -forest biomass
      /MAX BIOMASS FOREST) [(t ODM /ha)/Year]
deforestation = DEFORESTATIOIN RATE *(1 –STEP (1, FINAL YEAR
      DEFORESTATION)) * forest biomass [t ODM /(Year*ha)]
insect predation = MAX PREDATION RATE INSECTS * insect biomass
      *(forest biomass /(forest biomass + HALF SATURATION CONSTANT
      PREDATION BY INSECTS)) [t ODM /(Year*ha)]
forest biomass = INTEG (+increase forest – deforestation - insect predation,
      INITIAL FOREST BIOMASS) [t ODM /ha]
increase insects = REPRODUCTION RATE INSECTS * insect biomass
      *(1- insect biomass /(CAPACITY FACTOR INSECTS * forest biomass
      *(MAX BIOMASS FOREST - forest biomass))) [t ODM /(Year*ha)]
bird predation = MAX PREDATION RATE BIRDS * bird biomass *(insect biomass
      /(insect biomass + HALF SATURATION CONSTANT PREDATION BY BIRDS))
      [t ODM /(Year*ha)]
insect biomass = INTEG (+increase insects - bird predation, INITIAL INSECT
      BIOMASS) [t ODM /ha]
increase birds = REPRODUCTION RATE BIRDS * bird biomass *(1- bird biomass
      /(CAPACITY FACTOR BIRDS * forest biomass * insect biomass))
      [t ODM /(Year*ha)]
bird biomass = INTEG (+increase birds, INITIAL BIRD BIOMASS) [t ODM /ha]

Simulation time parameters
INITIAL TIME = 0 [Year]
FINAL TIME = 50 [Year]
TIME STEP = 0.01 [Year]

Simulation results

Figure Z406b shows the time plot of the three biomasses for the default parameter
setting. In this case the initial state corresponds to a forest fraction of 20 percent. At
first *forest biomass* still increases since the rate of *increase forests* is twice as large as
the rate of *deforestation*, and *insect predation* has only a minor effect. But *insect biomass* gradually increases to an order of magnitude which can no longer be checked by
the strongly increasing *bird biomass*. The remaining forest collapses quickly by *insect
predation*; at this time the insect and bird populations also disappear for lack of food.

Figure Z406c shows that despite unfavorable initial conditions, collapse of the
system can be avoided by a reduction of the DEFORESTATION RATE.

Figure Z406d shows simulation results for an initially larger forest fraction (INI-
TIAL FOREST BIOMASS = 30) and a higher DEFORESTATION RATE (= 0.1). In this case
the system finally collapses around year 30 after an explosion of the insect and bird
populations.

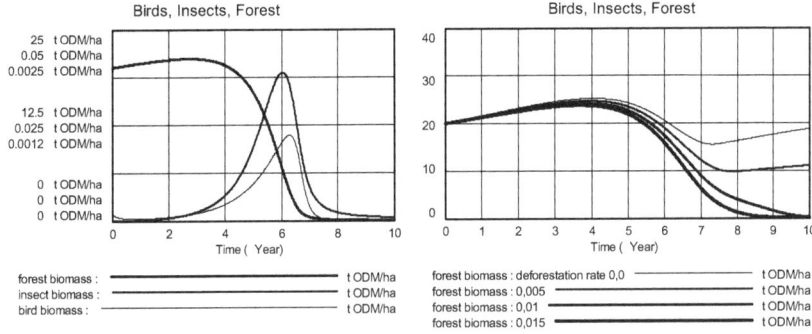

Figure Z406b: The ecosystem collapses if the forest fraction becomes too small.
Figure Z406c: Reduction of deforestation prevents ecosystem collapse.

In Figure Z406e it can be seen that (for otherwise identical conditions as in Figure Z406d) collapse is avoidable by ending deforestation in time (FINAL YEAR DEFORESTATION = 20). With *forest biomass* recovering quickly, *insect biomass* becomes insignificant, and *bird biomass* is reduced correspondingly also. Interestingly enough, regular oscillations develop for these two populations as long as *woods biomass* is relatively small.

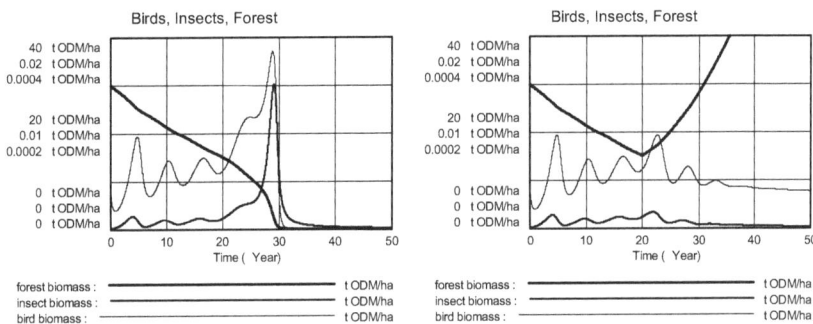

Figure Z406d: The populations of birds and insects oscillate before the final collapse due to deforestation.
Figure Z406e: Collapse is avoided if deforestation is stopped in time.

The simulation confirms the initial assumptions of the researchers: If the forest fraction remains large enough, birds and insects survive in small populations. If the forest fraction becomes too small, conditions for the insects improve very strongly and an explosive insect pest outbreak will occur which either completely destroys the remaining forest, or decimates it temporarily. The forest losses by deforestation determine decisively any further development and the possibility of collapse

Exercises

1. Examine the role of INITIAL FOREST BIOMASS and the consequences of different DEFORESTATION RATE scenarios. Under what conditions is collapse by insect predation unlikely?

2. In which parameter range are oscillations of the insect and bird populations to be expected without, however, leading to collapse of the ecosystem?

3. Determine whether by a timely pest control campaign the mass multiplication of pest insects and collapse of the forest can be prevented. (Add parameters EXTERMINATION FRACTION and TIME OF PEST CONTROL to the model). What fraction of the *insect biomass* has to be destroyed to prevent the collapse? What point in time is most effective?

References

Richter, O. 1985: *Simulation des Verhaltens ökologischer Systeme*. VCH, Weinheim, S. 91-98.

Trenbath, B. R., Smith, A. D. M. 1981: Basic concepts for a systems analysis of eucalypt dieback in New South Wales. In: K. M. Old, G. A. Kile, C. P. Ohmart (eds.): *Eucalypt dieback in forests and woodlands*. CSIRO, Australia, p. 234-243.

Z407 Plant competition and nutrient cycle

Simulation task

Ecosystems produce stand litter (leaf litter, deadwood, dead organisms). This litter is used energetically by soil organisms in often complex food-chains and decomposition processes. Nutrients contained in litter are released by mineralization and become again available for uptake by plants and for corresponding plant growth – the nutrient cycle is largely closed, and the ecosystem therefore largely fertilizes itself.

Soil nutrients are taken up simultaneously by all plants growing on it, by annuals (with their high mortality rates and high rates of nutrient uptake) as well as long-lasting plants like trees (with low mortality rates and low rates of nutrient uptake). The different plant species of an ecosystem with their different characteristics are in competition. It is not immediately obvious what consequences the coupling of the different processes of different species can have on the dynamic development of the ecosystem. A simulation model can assist in determining possible dynamics.

Plant growth of two plant populations is modeled as function of nutrient availability. Their common litterfall is mineralized. The nutrient contained in it (represented here as nitrogen N) is then available for uptake by the two plant populations. Growth and mineralization are primarily dependent on temperature; a seasonal effect on growth and mineralization is therefore implemented in the model. The amount of nitrogen per hectare [tN/ha] is used as the unit of measurement, and is also used to measure biomasses. To express biomass in terms of amount of carbon, the correct carbon-to-nitrogen ratio (C/N ratio) must be applied: about 20 for green biomass, and about 100 for wood. From the amount of carbon the amount of organic dry matter (ODM) can be computed: 1 t ODM contains about 0.45 t C. For fixing this carbon, plants have to take up $44/12 = 3.67$ as much CO_2 from the atmosphere, i.e. 1.65 t CO_2 per t ODM (Bossel 1990/1994, 42-68, see also the models for climate and forest growth in the present volume).

The model therefore describes on the one hand the nutrient cycling of a terrestrial ecosystem, on the other hand the competition of two plant species competing for nutrients on the same piece of land. A similar scheme also applies to aquatic ecosystems.

Simulation model

Figure Z407a shows the simulation diagram. The corresponding model equations are listed in the following.

The stands of *plant X* and *plant Y* grow according to their *nutrient uptake X* and *nutrient uptake Y*. This leads to (total) *nutrient uptake* from the pool of *nutrient in soil*. The two plant formations lose biomass (measured in terms of N) by *dieback X* (with MORTALITY RATE X) and *dieback Y* (with MORTALITY RATE Y). This common *litterfall* increases the mineralizable *stand litter*. *Mineralization* occurs at the MINERALIZATION RATE modified by its *seasonal dependence*. *Mineralization* increases the *nutrient in soil* and *nutrient saturation*. This is represented by a Michaelis-Menten formulation (cf. model Z111 in *System Zoo 1*) depending on SATURATION CONSTANT and changing with *seasonal dependence*. The seasonal fluctuations of *mineralization* (temperature effect) and of *nutrient uptake* during the growth period are simulated using a sine function with a period of one year.

Differences between the two plant populations *plant X* and *plant Y* can be introduced by specifying different NUTRIENT UPTAKE RATES and MORTALITY RATES for X and Y.

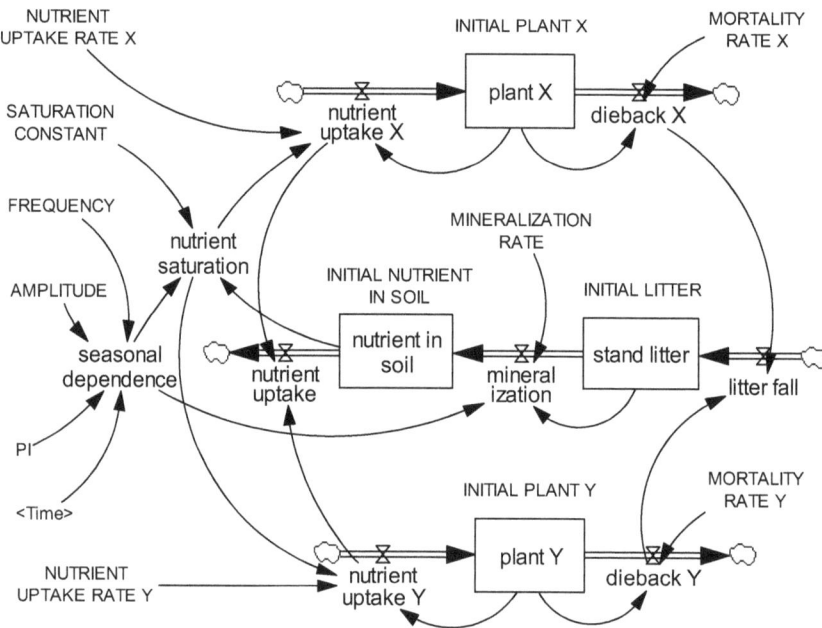

| Nutrient Cycle and Plant Competition |

Figure Z407a: Simulation diagram for plant competition and nutrient cycling.

Parameters and initial values
INITIAL LITTER = 0.1 [tN/ha]
INITIAL NUTRIENT IN SOIL = 1 [tN/ha]
INITIAL PLANT X = 0.1 [tN/ha]
INITIAL PLANT Y = 0.1 [tN/ha]
MINERALIZATION RATE = 0.5 [1/Year]
MORTALITY RATE X = 2 [1/Year]
MORTALITY RATE Y = 0.01 [1/Year]
NUTRIENT UPTAKE RATE X = 10 [1/Year]
NUTRIENT UPTAKE RATE Y = 1 [1/Year]
SATURATION CONSTANT = 0.5 [tN/ha]
FREQUENCY = 1 [1/Year]
AMPLITUDE = 1 [1]
PI = 3.14159 [1]

Dynamics

seasonal dependence = 1 + AMPLITUDE *sin (2 *PI * FREQUENCY *Time) [1]
nutrient saturation = (nutrient uptake /(SATURATION CONSTANT +nutrient in soil))
    *seasonal dependence [1]
nutrient uptake X = NUTRIENT UPTAKE RATE X *plant X *nutrient saturation
    [(tN/ha)/Year]
nutrient uptake Y = NUTRIENT UPTAKE RATE Y *plant Y *nutrient saturation
    [tN/(Year*ha)]
dieback X = MORTALITY RATE X *plant X [tN/(Year*ha)]
dieback Y = MORTALITY RATE Y *plant Y [tN/(Year*ha)]
plant X = INTEG (+nutrient uptake X -dieback X, INITIAL PLANT X) [tN/ha]
plant Y = INTEG (+nutrient uptake Y -dieback Y, INITIAL PLANT Y) [tN/ha]
litter fall = dieback X +dieback Y [tN/(Year*ha)]
mineralization = MINERALIZATION RATE *stand litter *seasonal dependence
    [tN/(Year*ha)]
nutrient uptake = nutrient uptake X +nutrient uptake Y [tN/(Year*ha)]
stand litter = INTEG (+litter fall -mineralization, INITIAL LITTER) [tN/ha]
nutrient in soil = INTEG (+mineralization -nutrient uptake, INITIAL NUTRIENT IN
    SOIL) [tN/ha]

Simulation time parameters

INITIAL TIME = 0 [Year]
FINAL TIME = 20 [Year]
TIME STEP = 0.01 [Year]

Simulation results

Figure Z407b shows the time plot for the default parameter setting. The initially available nutrient is quickly taken up by the plants and accumulated in their biomass. *Litter
fall* increases according to the specific MORTALITY RATES X and Y and refills the pool
of *nutrient in soil* by *mineralization*. Although initially *plant X* with its high NUTRIENT
UPTAKE RATE X can fully use the available *nutrient in soil* for rapid growth, it quickly
loses the accumulated biomass again because of its high MORTALITY RATE X. The
corresponding *litter fall* from *dieback X* soon contributes again by *mineralization* to
nutrient in soil. By contrast, the long-lived *plant Y* with its low MORTALITY RATE Y
can gradually build up permanent biomass. A growth limitation is provided by the
nutrient in the system, which now constantly recycles and maintains the biomass of
plant Y at a high level. *Plant X* disappears completely. Therefore in the long run the
plant population developing more permanent biomass and less stand litter dominates
(i.e. low MORTALITY RATE).

    The default parameter choice corresponds to a pioneer species *plant X* (*r*-
strategist) and a climax species *plant Y* (*k*-strategist). *Plant X* has a quick nutrient uptake and short lifetime (0.5 year); *plant Y* by contrast has a slow nutrient uptake and
long lifetime (100 years). The simulation shows that although the pioneer species
dominates at first, it is gradually overtaken by the climax species. The pioneer plant
finally disappears. Note that this competitive advantage arises exclusively from nutrient availability. Other differences that are relevant in real systems, like differences
with respect to photoactive radiation (shading of smaller plants) or water (root depth)
have not played a role here.

Figure Z407c shows the simulation result for the case that both plant populations have identical NUTRIENT UPTAKE RATE X and Y = 10, but slightly different MORTAL-ITY RATE X = 1 and MORTALITY RATE Y = 0.9. In this case *plant Y* with its lower MORTALITY RATE Y dominates in the long run.

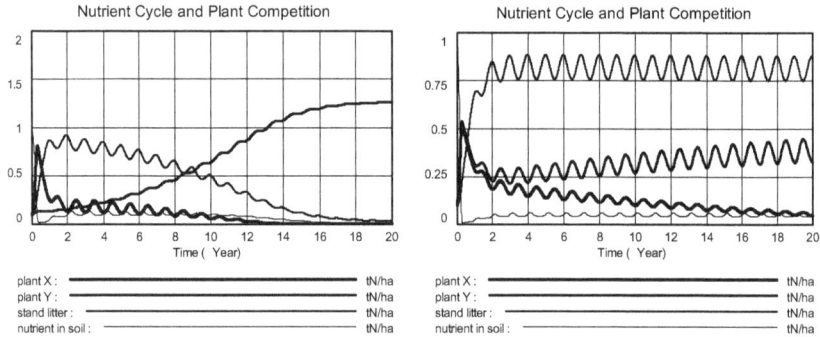

Figure Z407b: Development of two plant formations competing for the same nutrient pool. Plant X is short-lived (an annual), plant Y is long-lived (a tree).
Figure Z407c: Development of two competing (short-lived) plant populations. In the long run plant Y with somewhat more favorable parameters dominates.

Exercises

1. Examine the effect of different parameter assumptions for system development, particularly the parameters NUTRIENT UPTAKE RATE X and Y and MORTALITY RATE X and Y.
2. Try to adapt the model to competition of two real tree species – one a "pioneer" species, the other a "climax" species.

References

Bossel, H. 1990/1994: *Umweltwissen – Daten, Fakten, Zusammenhänge.* Springer, Berlin /Heidelberg /New York.
Jørgensen, S.E. 1992: Exergy and ecology. *Ecol. Modelling* 63, p. 185-214.

Z408 Fish pond

Simulation task

In their mature state ("climax") terrestrial or aquatic ecosystems can exist indefinitely because all mineral resources needed for life processes of the system are recycled. The material losses of such systems are small and they can therefore be replaced by natural processes with relatively small rates of production such as rock weathering or nitrogen fixation by free-living bacteria. Organic farming is based on this principle but it is also fundamental for pond agriculture.

Ponds integrated with other farming activities play a large role in agricultural systems in many regions of Asia (e.g. South China) – not only as producers of fish but also as effective decomposers of organic waste and mineralizers of nutrients and therefore as sources of fertilizer, producers of pig feed (water hyacinths), and reservoirs for irrigation waters.

If a system boundary is drawn around the pond, then the following system quantities can be recognized as major inputs to the fish pond system: solar radiation, precipitation, air temperature, water inflow, stocking of young fish, nutrients in the inflow, organic waste (of cattle, poultry, and humans), mineral fertilizers, and human labor. The main outputs of the pond system are: fish, nutrients in mud (fertilizer), organic substances in mud, water outflow, export of nutrients, export of sediment in outflow. Several different processes take place in the pond simultaneously: accumulation of sediment, decomposition of organic waste of every type, growth of algae and aquatic plants promoted by these nutrients, fish feeding on algae and aquatic plants, catching of fish as food, removal of mud for fertilization of fields, conservation of nutrients and soil particles washed out from fields into the pond. Nitrogen-fixing blue-green algae can supply an additional input of nitrogen.

The amount of algae and aquatic plants depends mainly on the amount of nutrients in the water, where the limiting nutrient (most often phosphate) plays the decisive role. Biomass in the pond also depends on the seasonal variation of water temperature and on solar radiation, which supplies the energy for photosynthesis and the formation of glucose. Both have a seasonal effect on the primary production of algae and plants. Nutrients in the pond recycle if there is no nutrient export (by removal of mud, aquatic plants, or fish). Nutrients in the water and in mud and sediment at the bottom of the pond are taken up by plants and algae and fixed in the organic matter produced by photosynthesis. Some of this material becomes litter, and is decomposed and mineralized again in the water or in the mud, setting the nutrients free again. Another part is consumed by fish or other plant-eating animals (e.g. ducks). Part of this is assimilated by the animals (net production); the major part either becomes waste because of feeding and digestion losses, or because the energy is used for respiration, while the nutrients are again set free in the animal feces. Again, the organic waste enters the decomposition and mineralization process, and the nutrients reenter the nutrient pool.

The management of fish ponds interferes with this closed nutrient cycle in various ways. First, there is an exportation of nutrients with each fish harvest, or with predation by water fowl or other animals. If not replaced by nutrients from the atmosphere, land runoff, stream inflow, or from mineral or organic fertilization, there would be gradual nutrient depletion in the pond ecosystem. Second, there is much greater nutrient loss by removal of mud for fertilization of fields. These nutrient losses are partially offset by the natural inputs just mentioned, but in an aquaculture system, the input of organic fertilizer, organic wastes, plant residues, or even mineral fertilizer, are much more significant. These nutrient inflows and outflows have to be managed in such a way that eutrophic (over-fertilized) and hypereutrophic conditions are avoided, while allowing plant and algae growth that produces maximum fish yield. Also, the system should produce the amount of fertilizer (mud) needed for fertilization of fields.

This management task is made difficult by the fact that we are dealing with several different state variables with very different typical time constants: algae and water plants grow much faster than fish, and microorganisms, the decomposers in the system, have even much faster turnover rates. We therefore must expect strong dynamic effects, complicated by the interactions of the different state variables. In fact, natural aquatic systems do show very fast dynamics – algal blooms and fish kills are examples. In the formulation of the model we concentrate on description of the interacting dynamics of these various processes to improve understanding of the fishpond system and its management within an ecofarm system.

This defines the purpose of the simulation model: Identification of important elements and important structural connections in the fishpond system, and construction of a dynamic model that provides an accurate description of the dynamic behavior of the state variables of the fishpond under various assumed management strategies. The purpose is better understanding of the dynamics of a fishpond, and their response to management inputs, not the exact prediction of time-dependent development in a particular pond.

The model computes algae as function of seasonal solar radiation and nutrient concentration in the pond. Algae are eaten by fish. Under eutrophic conditions algal blooms and corresponding algae die-off episodes can occur. The fish population depends on algae and aquatic plants as food source for its respiration needs and biomass growth. The organic wastes from the metabolism of fish or the dying of algae are decomposed with a decomposition rate that depends on seasonal water temperature. Nutrients are then available again in the water for uptake by algae growth. The pond dynamics is influenced by management measures, such as inputs of organic waste, of mineral fertilizer, of young fish as well as by the removal of fish for food, and mud for fertilization.

A more detailed description of the model is found in Bossel 1992.

Simulation model

The model structure is presented in the simulation diagrams of Figure Z408a and b. It contains the four state variables *algae*, *fish*, *organic waste* and *nutrient*. The corresponding model equations are listed in the following.

Algae: The state variable of this subsystem is the current stock (level) of *algae* biomass (organic dry substance). For the sake of model simplification, we aggregate into this *algae* compartment also phytoplankton, zooplankton, and water plants, inasmuch as they play a role in the food chain and nutrient cycle. The *algae growth rate* is determined primarily by *nutrient concentration* in the water which determines *net primary production algae*; it is modified by the seasonal change of *solar radiation*. This absolute growth rate as function of *nutrient concentration* and irradiation applies to a constant NORMAL ALGAE BIOMASS and must be modified therefore by *relative algae biomass:* If there are few *algae*, the *algae growth rate* will be much smaller than if algae concentration is higher.

Two loss rates have to be considered: First, *algae* are consumed by fish with an *algae consumption rate* corresponding to *food demand of fish* (and SPECIFIC FOOD CONSUMPTION OF FISH), and second, algae may die off because of adverse environmental conditions. Under extreme growth conditions (in algal blooms), the growth process itself may lead to a rapid dieback of the *algae* population as the growth conditions are negatively affected by the high *algae density* (nutrient depletion, toxic products of high decomposition rate, etc.). For this case a high *algae die-off rate* is introduced.

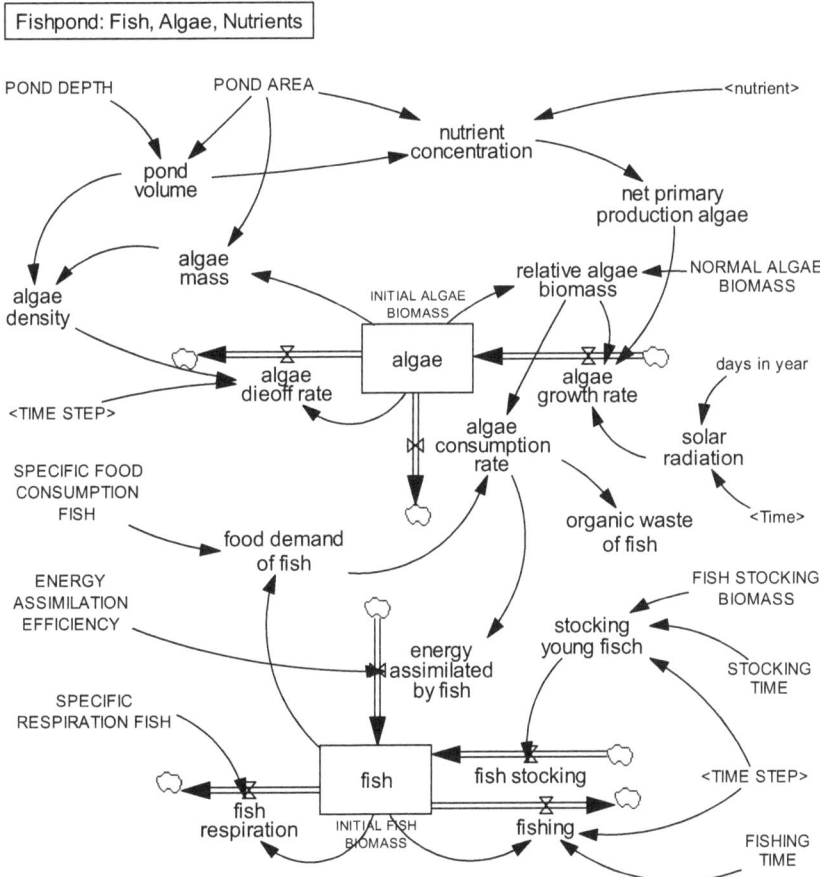

Figure Z408a: Simulation diagram for fish, algae, and nutrients, part 1.

Fish: The appropriate state variable is the current stock of *fish* (measured in organic dry matter of its biomass). Initially, there are no fish in the fishpond. At a given date in the spring (STOCKING TIME), the pond is stocked with a certain amount of young fish (FISH STOCKING BIOMASS). These fish then grow according to *energy assimilated by fish*, which corresponds to *algae consumption rate*, which is a function of *food demand of fish* and *relative algae biomass*. The organic dry matter of food or biomass is here used as a measure of energy, since the conversion factor from organic dry matter (ODM) to energy is fairly constant across all animal and plant species (about 17 kJ/g for carbohydrates, 21 kJ/g for proteins, 38 kJ/g for fat (lipoids), resulting in an average value of about 20 kJ/g for organic matter). The absolute food demand of fish is computed from the SPECIFIC FOOD CONSUMPTION FISH and the current biomass of the *fish* stock.

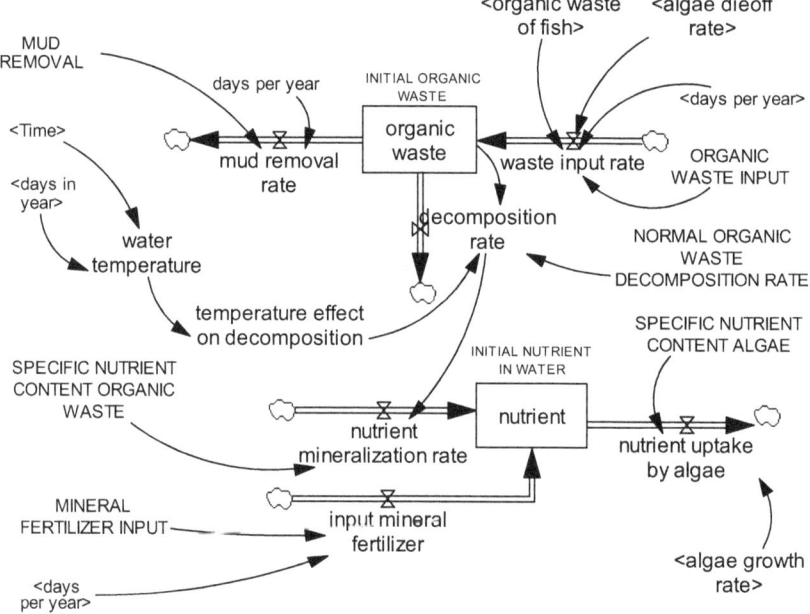

Figure Z408b: Simulation diagram for fish, algae, and nutrients, part 2.

With the assumption that the only food available to the *fish* is *algae* (and water plants also accounted for by this state variable), the *food demand of fish* translates into a corresponding *algae consumption rate*. However; if the amount of *algae* is not sufficient to satisfy the demand, the *algae consumption rate* is modified correspondingly in the model. All nutrients taken up by *fish* with their food (except those stored in fish biomass – this relatively small amount is neglected here) must be released after digestion and will enter the water as *organic waste of fish* (fish excrement). The *fish* will be able to assimilate only part of the energy contained in the food consumed by them; hence the food consumed must be multiplied by ENERGY ASSIMILATION EFFICIENCY to obtain *energy assimilated by fish*.

Some of the energy assimilated by the *fish* is used for *fish respiration*, fueling their continuous life processes. *Fish respiration* will be proportional to current *fish* biomass, and to SPECIFIC RESPIRATION FISH of the species. If the fish do not find enough food, respiration will drain their biomass, as some of the energy stored in the fish is consumed in the process. If the amount of *energy assimilated by fish* is greater than *fish respiration*, then *fish* biomass will be able to increase. Finally, at FISHING TIME in the fall, some or all of the fish will be taken out of the pond, thus reducing *fish* biomass.

Organic waste and *nutrient:* The nutrient cycle in the pond must be closed by de-composition of *organic waste*, and mineralization of nutrients contained in it. Since the net primary production in the pond is determined by the limiting nutrient, we only have to deal with that particular nutrient to compute the growth dynamics in the pond. Normally, the limiting nutrient in aquatic ecosystems is phosphorus. In our model we use conversion data for nitrogen, however, since this may actually be more significant in managed ponds. A conversion to phosphorus data is trivial, however, and it would not change the dynamics of the model.

Organic waste is decomposed by microorganisms with *decomposition rate*, re-leasing the *nutrient* into the water at a certain *nutrient mineralization rate*. Naturally, this process depends on amount and function of microorganisms present. In a more complex model, microorganisms should be represented by their own state variable. Here it was assumed that whenever organic wastes are present, the microorganism population will grow fast enough to adjust to any waste level, and that therefore the microorganism population does not have a controlling influence on the decomposition process. The *decomposition rate* is determined by amount of *organic waste*, NORMAL ORGANIC WASTE DECOMPOSITION RATE, and *temperature effect on decomposition* which depends on seasonal *time* and corresponding *water temperature*. The seasonal *water temperature* distribution is assumed to be sinusoidal, and to oscillate between the January and July extremes of 8 and 32 deg C, respectively (values for South China).

The amount of *organic waste* in the pond is changed by the rates of deposition from fish excretions (*organic waste of fish*) and dead algae (*algae dieoff rate*), by rate of ORGANIC WASTE INPUT from outside of the fishpond (grass, vegetable and crop wastes, wastes from livestock and human population, etc.), by *mud removal rate* for use as fertilizer on fields, and by *decomposition rate* of *organic waste*. It is convenient to again measure the amount of *organic waste* by its organic dry matter content (rep-resenting energy content, as explained above).

The amount of *nutrient* in the pond is increased by *nutrient mineralization rate* from decomposition of *organic waste* (with SPECIFIC NUTRIENT CONTENT ORGANIC WASTE), and by possible MINERAL FERTILIZER INPUT and resulting rate of *input min-eral fertilizer*. The *nutrient* pool is decreased by rate of *nutrient uptake by algae* (and water plants).

Important technical note: The model uses a single pulse to simulate *algae dieoff rate* for the case of algal bloom, producing sudden die-off of 90 percent of *algae*. This formulation is computed correctly with the Euler-Cauchy integration procedure. If an integration procedure with variable step size is used (such as the RK4 procedure inte-grated in Vensim PLE), incorrect results are obtained. This model must therefore be computed using the Euler-Cauchy procedure or a procedure with fixed step size

(TIMESTEP). In Bossel 1992 the algal bloom process with its drastically increased *algae die-off rate* is simulated more correctly by a die-off process lasting several days. However, formulation of this process using a graphic-interactive modeling procedure (as used here) turns out to be rather complicated. Euler-Cauchy integration using the simple formulation applied here leads to results which are almost identical to those of the original model.

Parameters and initial values
POND AREA = 1 [ha]
POND DEPTH = 2 [m]
INITIAL ALGAE BIOMASS = 100 [kg/ha]
INITIAL FISH BIOMASS = 0 [kg/ha]
INITIAL ORGANIC WASTE = 1 [kg/ha]
INITIAL NUTRIENT IN WATER = 50 [kgN/ha]
ENERGY ASSIMILATION EFFICIENCY = 0.6 [1]
NORMAL ALGAE BIOMASS = 1000 [kg/ha]
SPECIFIC NUTRIENT CONTENT ALGAE = 0.02 [kgN/kg]
SPECIFIC FOOD CONSUMPTION FISH = 0.1 [kg/(kg*Day)]
SPECIFIC RESPIRATION FISH = 0.04 [kg/(kg*Day)]
SPECIFIC NUTRIENT CONTENT ORGANIC WASTE = 0.02 [kgN/kg]
NORMAL ORGANIC WASTE DECOMPOSITION RATE = 0.03 [1/Day]
days in year = 365 [Day]
days per year = 365 [Day/year]

Scenario parameters
FISH STOCKING BIOMASS = 100 [kg/ha]
STOCKING TIME = 10 [Day]
FISHING TIME = 3000 [Day]
MINERAL FERTILIZER INPUT = 0 [kgN/(ha*year)]
ORGANIC WASTE INPUT = 0 [kg/(ha*year)]
MUD REMOVAL = 0 [kg/(ha*year)]

Algae dynamics
pond volume = POND AREA *POND DEPTH [ha*m]
solar radiation = 1+0.6 *SIN (6.28 *Time /days in year -3.14/2) [1] *relative insolation*
nutrient concentration = (nutrient/pond volume) *POND AREA [kgN/(ha*m)]
        *10 kgN/(ha*m) = 1 mgN/ liter*
net primary production algae = WITH LOOKUP (nutrient concentration, ([(0, 0) -(600, 400)], (0, 0), (10, 10), (50, 15), (100, 20), (150, 20), (500, 20))) [kg/(Day*ha)]
        net primary production of algae
algae growth rate = solar radiation *net primary production algae *10 *relative algae biomass [kg/(Day*ha)]
algae dieoff rate = IF THEN ELSE (algae density > 1000, 0.9*algae /TIME STEP, 0) [kg/(ha*Day)] *for high algae density: algae bloom and die-off of 90 pct of algae*
algae consumption rate = IF THEN ELSE (relative algae biomass < 1, food demand of fish *relative algae biomass, food demand of fish) [kg/(Day*ha)]
organic waste of fish = algae consumption rate [kg/(Day*ha)]
algae = INTEG (+algae growth rate −algae dieoff rate -algae consumption rate, INITIAL ALGAE BIOMASS) [kg/ha]
algae mass = POND AREA *algae [kg]
relative algae biomass = algae /NORMAL ALGAE BIOMASS [1]
algae density = (algae mass /pond volume) [kg/(ha*m)]

Fish dynamics
food demand of fish = SPECIFIC FOOD CONSUMPTION FISH *fish [kg/(Day*ha)]
energy assimilated by fish = algae consumption rate *ENERGY ASSIMILATION
        EFFICIENCY [kg/(Day*ha)]
fish respiration = SPECIFIC RESPIRATION FISH *fish [kg/(Day*ha)]
stocking young fisch = PULSE (STOCKING TIME, TIME STEP) *FISH STOCKING
        BIOMASS /TIME STEP [kg/(Day*ha)]
fish stocking = stocking young fisch [kg/(Day*ha)]
fishing = PULSE (FISHING TIME, TIME STEP) *fish /TIME STEP [kg/(Day*ha)]
fish = INTEG (fish stocking +energy assimilated by fish −fishing -fish respiration,
        INITIAL FISH BIOMASS) [kg/ha]

Waste dynamics
waste input rate = organic waste of fish +ORGANIC WASTE INPUT /days per year
        +algae dieoff rate [kg/(Day*ha)]
mud removal rate = MUD REMOVAL /days per year [kg/(Day*ha)]
water temperature = 20 +12 *SIN (6.28 *Time /days in year -3.14/2) [Celsius]
temperature effect on decomposition = WITH LOOKUP (water temperature,
        ([(0, 0) -(60, 5)], (0, 0.05), (10, 0.25), (20, 1), (30, 1.5), (40, 2), (50, 2))) [1]
decomposition rate = temperature effect on decomposition *NORMAL ORGANIC
        WASTE DECOMPOSITION RATE *organic waste [kg/(Day*ha)]
organic waste = INTEG (waste input rate −mud removal rate -decomposition rate,
        INITIAL ORGANIC WASTE) [kg/ha]

Nutrient dynamics
input mineral fertilizer = MINERAL FERTILIZER INPUT /days per year [kgN/(Day*ha)]
nutrient mineralization rate = decomposition rate *SPECIFIC NUTRIENT CONTENT
        ORGANIC WASTE [kgN/(ha*Day)]
nutrient uptake by algae = algae growth rate *SPECIFIC NUTRIENT CONTENT
        ALGAE [kgN/(Day*ha)]
nutrient = INTEG (input mineral fertilizer +nutrient mineralization rate -nutrient uptake
        by algae, INITIAL NUTRIENT IN WATER) [kgN/ha]

Simulation time parameters
INITIAL TIME = 0 [Day]
FINAL TIME = 1825 [Day]
TIME STEP = 0.125 [Day]

Simulation results

Figures Z408c to e show time plots of simulation results for three consecutive years
calculated with the parameter values of the default set. Continuation of the simulation
for further years produces results which are completely identical with the results of the
third year (day 730-1095) and repeat every year. Obviously, by the third year a seasonal ecological cycle has evolved where *nutrients* are recycled and the dynamics of
algae, fish, organic waste and *nutrient* repeat identically every year.

During several algal blooms in the nutrient-rich water of the first year the stock
of *fish* (introduced into the pond on the 10th day) grows only gradually and reaches its
maximum only in early winter, but then declines strongly because of food shortage in
the winter season. (For this reason fish ponds are usually fished completely in late fall;
this is not implemented in this simulation run.) After the first year a larger pool of

organic waste (mud) has formed that now releases *nutrient* at a relatively steady rate, leading in following years to rapid growth of *algae* and water plants in spring, which offers sufficient food for the *fish* until summer. After that the stock of *algae* decreases again; correspondingly the stock of *fish* also declines gradually, reaching its minimum in winter.

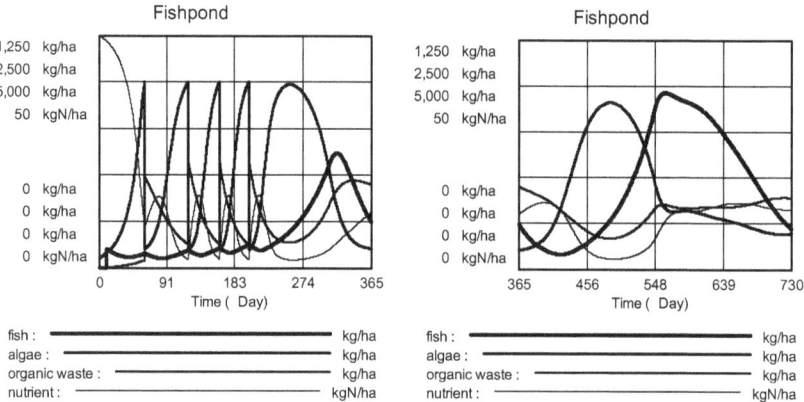

Figure Z408c: Development in the first year: Because of nutrient-rich water several algal blooms occur.
Figure Z408d: Development in the second year: Nutrients are now largely fixed in the mud; algal blooms do not occur any more.

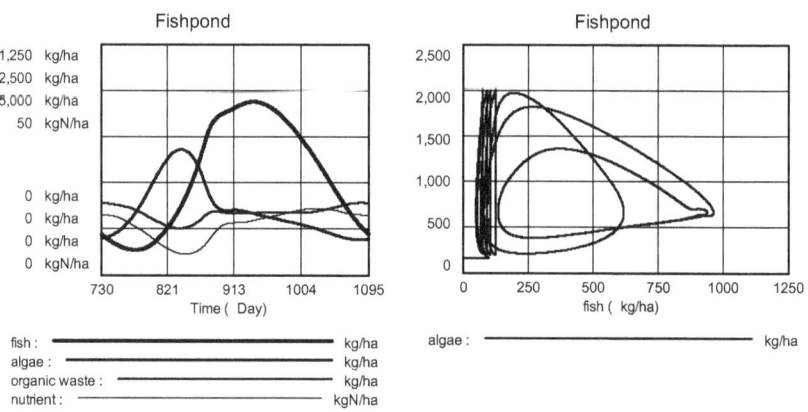

Figure Z408e: In the third year a stable cycle has evolved, which from now on repeats every year.
Figure Z408f: State diagram of long-term development: A permanently recurring cycle is established.

This sequence can also be recognized in the state diagram for *algae* and *fish* in Figure Z408f. After the algal blooms in the first year (spikes on the left) the state trajectory changes into a periodical oscillation which repeats identically after the third year. In spring *algae* population grows rapidly to a maximum of approximately 1400 kg/ha, while *fish* population reaches a peak of about 950 kg/hectares in late summer. The *organic waste* (mud) remains more or less constant at approximately 1500 kg/ha during the whole year, while the *nutrient* pool, with an average value of about 13 kgN/ha during most of the year, decreases to about 4 kgN/ha at the time of maximum algae growth in spring.

The simulation results of this simple model show behavioral trends, but the results hardly suffice for planning of management measures (like fish stocking, fish harvest, mud removal, fertilization etc.) for concrete applications. There are many improvements which can be introduced to the model to make it more accurate, such as the following:

- Nutrient storage in fish biomass is not taken into account now. This introduces an error when fish are harvested.
- A more accurate simulation of algae dieback following algal bloom should be developed.
- The effect of algal blooms on oxygen content of water is not considered. However, this is an essential determinant in algal blooms and subsequent fish kills.
- Description of the decomposition process is rather crude. However, it is unlikely that more detailed description will lead to very different results.
- Parameters and functions used now are mostly based on estimates and can be improved by more exact empirical values.

Exercises

1. Increase the initial *nutrient* pool gradually, and observe how it affects *algae* stock. At which nutrient level do algal blooms appear in later years also?

2. Try to develop, by repeated simulations with different parameters, a management strategy for maximum annual fish yield. For this, change INITIAL NUTRIENT IN WATER, FISH STOCKING BIOMASS, STOCKING TIME, and FISHING TIME.

3. Find initial settings (for January 1st) for the four state variables which lead to stable ecological cycles of different productivity (without removal of fish). For this the simulation period must extend over several years.

4. Find management strategies which include inputs of organic waste and mud removal at the end of the year and provide highest possible fish yield while allowing maximum removal of mud for fertilization.

5. Return an initially eutrophic (over-fertilized) pond to oligotrophic (nutrient-poor) state (no longer allowing algal blooms) by mud removal. How much mud per hectare will have to be removed for the initial parameters chosen by you?

6. Start with an initial oligotrophic state (*nutrient* content 10 kgN/ha). Increase the *nutrient* content by MINERAL FERTILIZER INPUT and try to obtain maximum fish yield this way. Describe your "experience".

7. Change the submodel for *fish* population by introducing feeding of fish. Try to achieve a maximum fish yield under oligotrophic conditions (no algae blooms) by feeding. What yields can you expect approximately?

8. Modify the calculation of *algae dieoff rate* as follows (cf. Bossel 1992: 229-230): If a critical *algae density* (= 1000, as now) is exceeded, a specific dieoff rate of 0.4 [1/day] is activated for 4 days, i.e. *algae dieoff rate* = 0.4 **algae*. Thereafter, *algae dieoff rate* = 0 again.

References

Bossel, H. 1992: *Simulation dynamischer Systeme – Grundwissen, Methoden, Programme*. Vieweg, Braunschweig und Wiesbaden, 2. Aufl., S. 218-244.

Z409 Fishery dynamics

Simulation task

Fishing is a classic example for use of a renewable resource. Unless overfished, fish populations regenerate. If regeneration is hardly impaired by fishing, then the fish population will persist at a constant size corresponding to its specific ecological environment. If the stock is overfished, the juvenile generation becomes too small to fully replace the adult generation. If overfishing continues, the population cannot recover and will collapse in short time. Even if fish catch stops now, it could take decades until the fish population recovers to its original size – if it hasn't become extinct meanwhile. In many regions of the world overfishing has led, and still leads, to the complete collapse of formerly huge fish populations: herring in the North Sea, codfish in the Northern Atlantic, tuna, whales – to name only a few. With the collapse of fish stocks came the collapse of the fishing industry in many regions. Employment and incomes disappeared; whole regions (like Newfoundland) lost their economic base.

Higher catches promise higher profits in the short run. However, this short-term economic interest of fishermen is not consistent with the need for maintaining the natural regeneration ability of fish populations to avoid losing stocks completely in the long run. To safeguard the sustainability of fishing, measures like closed seasons, minimum sizes, restrictions with respect to catch technology, catch limits, and fleet limitations are being applied. Apart from effective control of these measures, however, their proper design is difficult since the economic interests of fishermen and consumers must be optimally represented while fish stocks must be protected at a level which assures high and sustainable catch yields. Here again, simulation models can help, but the acquisition of sufficiently reliable data for their parameterization is often difficult.

In the following model it is assumed that in a large inland lake a single fish population is harvested. The complete fishery system consists of the coupled dynamic systems of the fish population on the one hand and the fishing industry on the other (represented by the investments in fishing boats), which uses this stock and depends on it. Earnings of the fishermen are used to buy new fishing boats to replace old boats that go out of commission and/or to enlarge the fleet.

On first reflection (taking the point of view of the fishermen) the fish catch and corresponding profit will be small if either too few or else too many boats go out with the fishing fleet. In the first case, the available catch potential is not fully used, in the second the cost of boats and their operation is too high relative to the income from a limited (and probably diminishing) fish stock. So there will probably be an optimum for the number of boats which, however, will depend on economic conditions (fish price and boat expenses) and on ecological conditions (sustainable fish catch). For the fishermen it would be important to know these conditions to regulate fishing in such a way that (1) neither an ecological collapse (of the fish population) nor an economic collapse (of the fishery enterprises) will occur, and (2) a lasting (sustainable) exploitation of the inland lake under optimal economic conditions is achieved. Since we are dealing with relatively complex ecological and economic processes that are intimately connected with each other, the development of a mathematical model for computer simulation under different assumed conditions and for the search for an optimal solution is in order. The step-by-step development of the following model is presented in full detail elsewhere (Bossel 1994a, 2004a). It also is shown there (and in Bossel 2007 *Systems and Models*) that its structure corresponds exactly to that of the traditional predator-prey system for limited carrying capacity (i.e. model Z402).

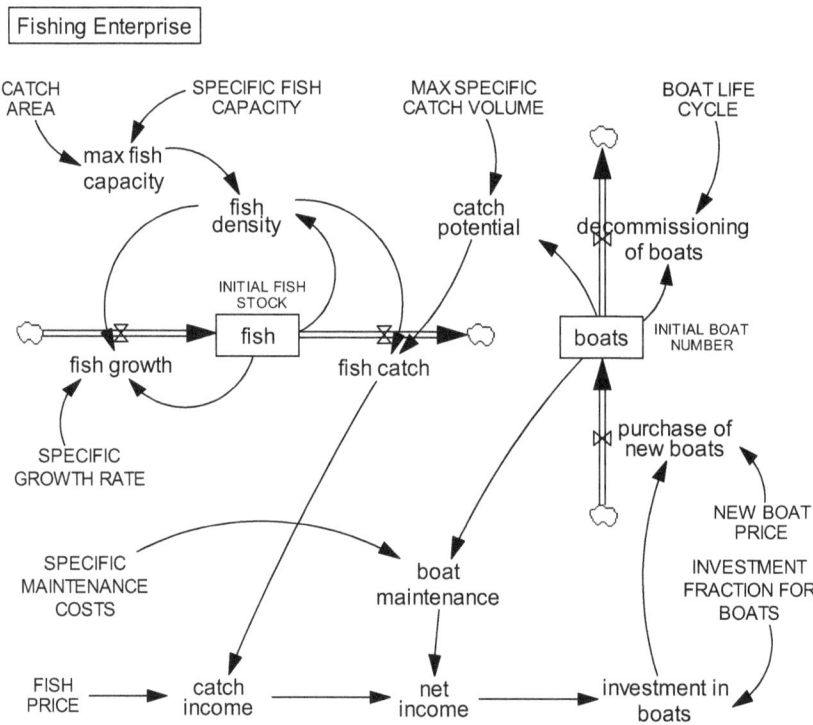

Figure Z409a: Simulation diagram for the fishing system

Simulation model

Figure Z409a shows the simulation diagram of the fishing model. The model equations are listed in the following.

For the submodel of the *fish population* we develop relationships 1 through 9 in the following list; they describe how *fish* population changes as a function of natural growth conditions and "predation" by fishermen. For the submodel of the *fishing fleet* we define relationships 10 through 27; they describe how *boat* number changes in response to earnings from fishing.

1. Net growth of the *fish* population (*fish growth*) depends on the current population of *fish*.
2. *Fish growth* is higher if the SPECIFIC GROWTH RATE of the *fish* population is higher.
3. *Fish growth* is reduced as *fish density* approaches a critical limit given by *max fish capacity*.

4. *Fish* population is increased by *fish growth*.
5. *Fish* population is reduced by *fish catch* (annual rate).
6. *Fish density* increases as *fish* population increases.
7. If *max fish capacity* is lower, relative *fish density* will be higher for the same *fish* population.
8. *Max fish capacity* depends on CATCH AREA of the lake.
9. *Max fish capacity* depends on SPECIFIC FISH CAPACITY of the lake. If there is no *fish catch*, *fish* population can grow up to its *max fish capacity* limit which is given by ecological conditions. If *fish* population approaches this limit, its annual net growth will eventually be reduced to zero.
10. Number of *boats* increases by annual *purchase of new boats*.
11. Number of *boats* decreases by annual *decommissioning of boats*.
12. Maximum possible annual fish catch of the fleet (*catch potential*) is determined by number of active *boats*.
13. *Catch potential* depends on boat performance (annual MAXIMUM SPECIFIC CATCH rate per boat).
14. Actual annual *fish catch* increases with *catch potential*.
15. *Fish catch* increases with *fish density*.
16. Annual *catch income* increases with *fish catch*.
17. *Catch income* is proportional to FISH PRICE.
18. *Net income* is higher if *catch income* is higher.
19. *Net income* is lower if *boat maintenance* costs are higher.
20. *Boat maintenance* costs are higher if SPECIFIC MAINTENANCE COSTS are higher.
21. *Boat maintenance* costs increase with number of *boats*.
22. If *net income* is higher, more capital is available for *investment in boats*.
23. If the fraction of net income reserved for investment in new *boats* (INVESTMENT FRACTION FOR BOATS) is higher, more capital is available for *investment in boats*.
24. If more *investment in boats* is possible, the rate of *purchase of new boats* will be greater.
25. If NEW BOAT PRICE is higher, the rate of *purchase of new boats* will be smaller.
26. Annual rate of *decommissioning of boats* depends on average BOAT LIFE CYCLE.
27. Annual rate of *decommissioning of boats* depends on current number of *boats*.

The 27 influence arrows in the model diagram of Figure Z409a correspond to the 27 relationships listed here.

Parameters and initial values
INITIAL FISH STOCK = 5000 [t fish]
CATCH AREA = 100 [km²]
SPECIFIC FISH CAPACITY = 100 [t fish /km²]
SPECIFIC GROWTH RATE = 1 [1/Year]
INITIAL BOAT NUMBER = 100 [boat]
BOAT LIFE CYCLE = 15 [Year]
NEW BOAT PRICE = 100000 [$/boat]
INVESTMENT FRACTION FOR BOATS = 0.5 [1]
MAX SPECIFIC CATCH VOLUME = 100 [t fish /(boat *Year)]
SPECIFIC MAINTENANCE COSTS = 50000 [$/(boat *Year)]
FISH PRICE = 1000 [$/t fish]

Dynamics
max fish capacity = CATCH AREA *SPECIFIC FISH CAPACITY [t fish]
fish growth = SPECIFIC GROWTH RATE *fish *(1 -fish density) [t fish /Year]
fish catch = catch potential *fish density [t fish /Year]
fish = INTEG (+fish growth -fish catch, INITIAL FISH STOCK) [t fish]
fish density = fish /max fish capacity [1]
purchase of new boats = investment in boats /NEW BOAT PRICE [boat /Year]
decommissioning of boats = boats /BOAT LIFE CYCLE [boat /Year]
boats = INTEG (purchase of new boats -decommissioning of boats, INITIAL BOAT
       NUMBER) [boat]
catch potential = MAX SPECIFIC CATCH VOLUME *boats [t fish /Year]
boat maintenance = SPECIFIC MAINTENANCE COSTS *boats [$/Year]
catch income = FISH PRICE *fish catch [$/Year]
net income = catch income -boat maintenance [$/Year]
investment in boats = INVESTMENT FRACTION FOR BOATS *net income [$/Year]

Simulation time parameters
INITIAL TIME = 0 [Year]
FINAL TIME = 20 [Year]
TIME STEP = 0.02 [Year]

Simulation results

Figure Z409b shows the time plots of *fish, boats* and *fish catch* for the default parameter setting. Initially the *fish* stock decreases strongly from an initial value of 5000 [t
fish] to a minimum of 3000, but then recovers and stabilizes at about 6300. The number of *boats* quickly drops from its initial value of 100 to about 36, where it stabilizes.
The annual *fish catch* reduces from its initial value of 5000 [t fish/year] to a minimum
of 1800 and then settles at about 2300. Obviously the system stabilizes after about 10
years in a dynamic equilibrium, where the state variables *fish* and *boats* no longer
change, and losses (by *fish catch* and *decommissioning of boats*) are exactly balanced
by gains (*fish growth* and *purchase of new boats*).

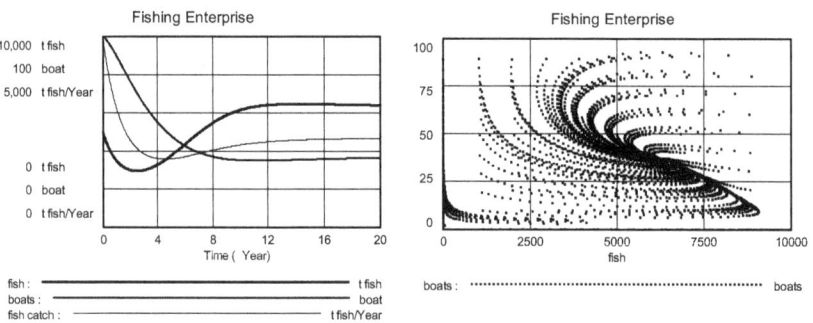

Figure Z409b: The system drifts towards an equilibrium state.
igure Z409c: Independently of the initial state, all state trajectories move towards the
same equilibrium state.

Figure Z409c shows the state diagram for *boats* as function of *fish*. This diagram was generated by coupling the fishery model to module Z115 "State diagram" (in *System Zoo 1*) and generating state trajectories for 100 different initial states of the system. This system also shows clearly that for all initial conditions in the domain examined, the trajectories move towards a stable equilibrium point (at about 37 boats and 6300 t of fish) (cf. Figure Z402Ac for the predator-prey system with limited prey). The results of further simulations with this model are documented in Bossel 1994 *Modeling and Simulation*.

Exercises

1. Investigate the behavior for changed parameters (in particular: SPECIFIC GROWTH RATE, MAX SPECIFIC CATCH VOLUME, NEW BOAT PRICE, INVESTMENT FRACTION FOR BOATS, FISH PRICE). Produce time plots of interesting cases for *fish*, *boats*, *fish catch*, *net income* (all variables in one plot) as well as state diagrams (*boats* vs. *fish*). Find the respective equilibrium points from these diagrams. In particular, investigate whether the system can collapse (at high catch rates).
2. Couple module Z115 "State diagram" to model Z409 "Fishery" and generate plots of state trajectories (like Figure Z409c) for interesting parameter combinations (cf. Exercise 1).
3. Introduce single-letter variables and parameters, and use mathematical substitution to reduce the model equations to a set of two differential equations of the predator-prey type (cf. model Z402A). Determine the (3!) equilibrium points – as mathematical expressions involving the system parameters – by applying the condition $dz/dt = 0$. Calculate the equilibrium points numerically after realistic quantification of system parameters (cf. Bossel 1994 *Modeling and Simulation*, 208-210).
4. Linearize the system of differential equations (using the Jacobian matrix) at the stable equilibrium point corresponding to the default parameters (or another interesting parameter combination). Find the characteristic equation of the linear substitute system and determine its eigenvalues. Draw the location of the eigenvalues in the complex number plane and use it to discuss the behavior of the model system near the equilibrium point. Is the behavior there stable? Are there oscillations? Are they damped? (cf. Bossel 1994 *Modeling and Simulation*, Ch. 7; Bossel 2007 *Systems and Models*, Ch. 2.8).
5. Determine the location of equilibrium points and the (sustainable) *fish catch* for different values of MAX SPECIFIC CATCH VOLUME in the range from 50 to 250. Plot the result as function of MAX SPECIFIC CATCH VOLUME and formulate fishing policy recommendations for optimal and sustainable exploitation of the fishing grounds.

References

Bossel, H. SDS 2004: *Systeme, Dynamik, Simulation – Modellbildung, Analyse und Simulation komplexer Systeme*. Books on Demand, Norderstedt, S. 202-225.

Z410 Fishery with optimization

Simulation task

Simulations with model Z409 "Fishery" will occasionally yield a collapse of the boat fleet (under unfavorable economic conditions) but never a complete collapse of the fish population. The explanation for this is found in the model equations: The *fish catch* depends on *fish density*

$$\textit{fish catch} = \textit{catch potential} \cdot \textit{fish density}$$

This means that for decreasing *fish density* (and correspondingly decreasing *fish* stock) the *fish catch* ultimately goes to zero. This has corresponding economic consequences for the fishery enterprises, however: The number of *boats* also decreases strongly, reducing the *fish catch* further. The *fish* population is thus protected from complete collapse. Finally equilibrium always develops between *fish* population and the number of *boats*.

The implicit assumption used here that the *fish* are evenly distributed over the catch area does not correspond to reality in the case of many fish species of economic interest. These often appear in shoals of fish which can easily be located with sonar technology. This means that under these conditions the *fish catch* does not depend any more on (average) *fish density* but only on the quality of locator technology, (i.e. the CATCH CHANCE) and the *catch potential* of the fleet:

$$\textit{fish catch} = \textit{catch potential} \cdot \text{CATCH CHANCE}$$

As a consequence, the system behavior changes fundamentally, however.

To modify the fishery model correspondingly, we first introduce three new parameters concerning use of FISHFINDER TECHNOLOGY, authorized MAX BOAT NUMBER, as well as CATCH CHANCE that takes into account that even with sophisticated locator technology the *fish catch* is smaller than the available *catch potential*.

To use the model in the search for an optimal management strategy, a control criterion for guiding the search must be introduced. Two points of view in particular can play a decisive role in this search: (1) maximizing of *fish catch* (without consideration of cost) or (2) maximizing of *profit rate* (without consideration of the amount of *fish catch*). In reality both will have to be taken into account. In the model we use a *quality index* which takes into account both points of view, each with a respective weighting selected by the model user.

A more detailed representation of the modified model and its results is found elsewhere (Bossel 1994 *Modeling and Simulation*: 264-273).

Simulation model

Figure Z410a shows the simulation diagram. The corresponding model instructions are listed in the following. Model Z409 "Fishery" was augmented by the parameters FISHFINDER TECHNOLOGY, CATCH CHANCE, and MAX BOAT NUMBER, by reformulating the instructions for *fish catch* and *purchase of new boats*, and by adding instructions for computation of *quality* as function of the results for *relative profit rate* and *relative catch volume*, which are weighted by PROFIT WEIGHT and CATCH VOLUME WEIGHT, respectively.

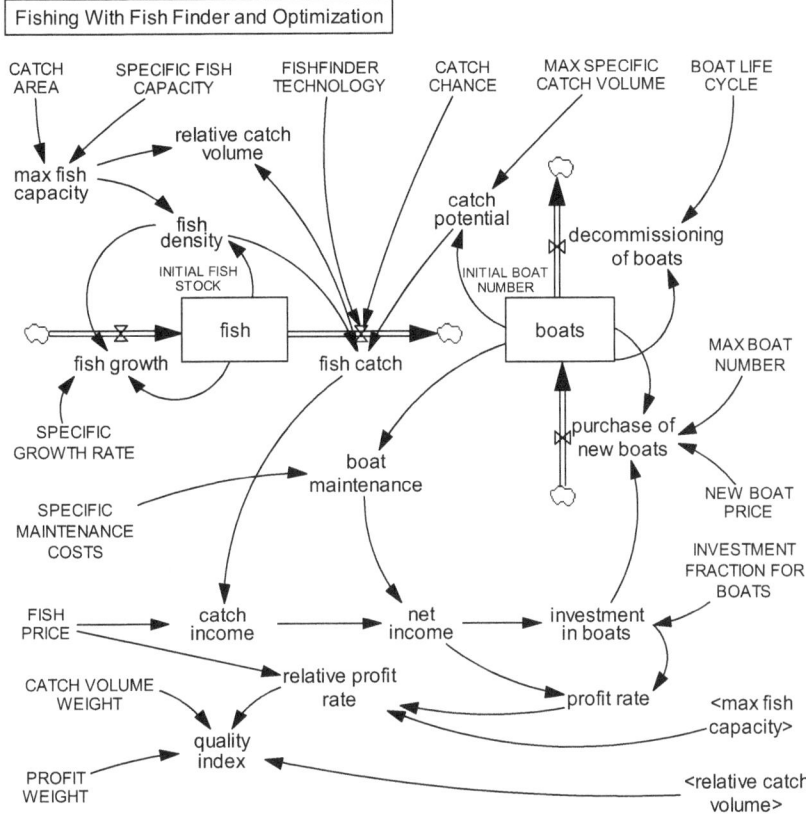

Figure Z410a: Simulation diagram for fishery with optimization.

Parameters and initial values
INITIAL FISH STOCK = 5000 [t fish]
CATCH AREA = 100 [km²]
SPECIFIC FISH CAPACITY = 100 [t fish /km²]
SPECIFIC GROWTH RATE = 1 [1/Year]
INITIAL BOAT NUMBER = 10 [boat]
MAX BOOTSZAHL = 25 [boat]
BOAT LIFE CYCLE = 15 [Year]
NEW BOAT PRICE = 100000 [$/boat]
INVESTMENT FRACTION FOR BOATS = 0.5 [1]
FISHFINDER TECHNOLOGY = 1 [1] *available = 1, not available = 0*
CATCH CHANCE = 0.8 [1]
MAX SPECIFIC CATCH VOLUME = 100 [t fish /(boat *Year)]
SPECIFIC MAINTENANCE COSTS = 50000 [$/(boat *Year)]
FISH PRICE = 1000 [$/t fish]
CATCH VOLUME WEIGHT = 0 [1]
PROFIT WEIGHT = 1 [1]

Dynamics
max fish capacity = CATCH AREA *SPECIFIC FISH CAPACITY [t fish]
fish growth = SPECIFIC GROWTH RATE *fish *(1 -fish density) [t fish /Year]
fish catch = IF THEN ELSE (FISHFINDER TECHNOLOGY = 0, catch potential *fish
        density, IF THEN ELSE (fish density > 0, catch potential *CATCH CHANCE, 0))
        [t fish /Year]
relative catch volume = fish catch /max fish capacity [1/Year]
fish = INTEG (+fish growth -fish catch, INITIAL FISH STOCK) [t fish]
fish density = fish /max fish capacity [1]
purchase of new boats = IF THEN ELSE (boats > MAX BOOTSZAHL, 0, investment in
        boats /NEW BOAT PRICE) [boat /Year]
decommissioning of boats = boats /BOAT LIFE CYCLE [boat /Year]
boats = INTEG (purchase of new boats -decommissioning of boats, INITIAL BOAT
        NUMBER) [boat]
catch potential = MAX SPECIFIC CATCH VOLUME *boats [t fish /Year]
boat maintenance = SPECIFIC MAINTENANCE COSTS *boats [$/Year]
catch income = FISH PRICE *fish catch [$/Year]
net income = catch income -boat maintenance [$/Year]
investment in boats = INVESTMENT FRACTION FOR BOATS *net income [$/Year]
profit rate = net income -investment in boats [$/Year]
relative profit rate = profit rate /(FISH PRICE *max fish capacity) [1/Year]
quality index = ((CATCH VOLUME WEIGHT *relative catch volume) +(PROFIT
        WEIGHT *relative profit rate)) *100/(CATCH VOLUME WEIGHT +PROFIT
        WEIGHT) [1/Year]

Simulation time parameters
INITIAL TIME = 0 [Year]
FINAL TIME = 20 [Year]
TIME STEP = 0.02 [Year]

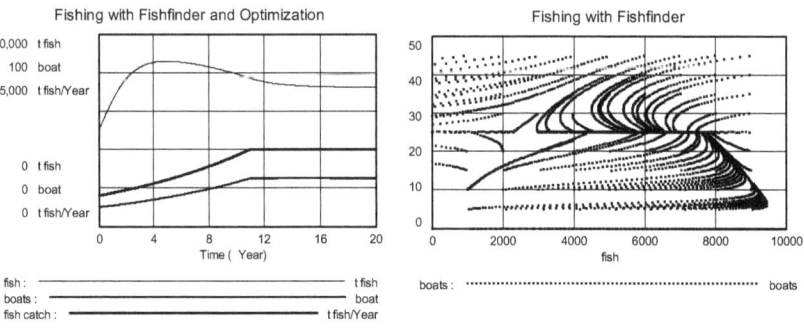

Figure Z410b: An equilibrium state appears if the boat number is limited to 25.
Figure Z410c: If fish locator technology is used, the system is in danger of collapse.

Simulation results

Figure Z410b shows the time response of the model for the default parameter settings.
To generate stable behavior the initial value for *boats* must be low, and MAX BOAT

NUMBER must be restricted to a small value (here 25).

The state diagram of Figure Z410c, generated by coupling the model to module Z115, clearly shows that system behavior has changed completely compared to model Z409 as a result of using fish locator technology. A stable equilibrium state exists only if the limitation MAX BOAT NUMBER is introduced. But also in this case the system collapses if initially the ratio of *boats* to *fish* is too large (left hand region in state diagram Z410c). In the stable region the state trajectories move towards an equilibrium point of *boats* = 25 and *fish* = 7236. In contrast to linear dynamic systems we find different stability conditions in the different state regions, a common quality of nonlinear systems.

A fishery enterprise operating under the tenet of economic profitability would have to deal with the important question of what share of annual profits (here: *net income*) should be reinvested into purchase *of new boats* (INVESTMENT FRACTION FOR BOATS) under conditions of equilibrium. A large number of *boats* means high costs for maintenance and operation (*boat maintenance*), and would cut into the higher profit that a larger boat fleet could produce, while too small a *boat* fleet can produce only small *fish catch* and corresponding low *profit rate*. The result of such deliberations depends entirely on the weightings assigned to the different decision criteria.

Figures Z410d and e show the *quality* index for five simulation runs where the INVESTMENT FRACTION FOR BOATS was varied between 0.1 and 0.9. In the first case (Figure Z410d) CATCH VOLUME WEIGHT = 1 and PROFIT WEIGHT = 5 were used, i.e. maximization of profit was the dominant criterion. An optimal result is obtained for an INVESTMENT FRACTION FOR BOATS = 0.5. In the second case (Figure Z410e) maximization of *fish catch* was the dominant concern, with CATCH VOLUME WEIGHT = 5, and PROFIT WEIGHT = 1. In this case, the optimum INVESTMENT FRACTION FOR BOATS = 0.9, resulting in rapid expansion of the fishing fleet to the permitted number (MAX BOAT NUMBER = 25).

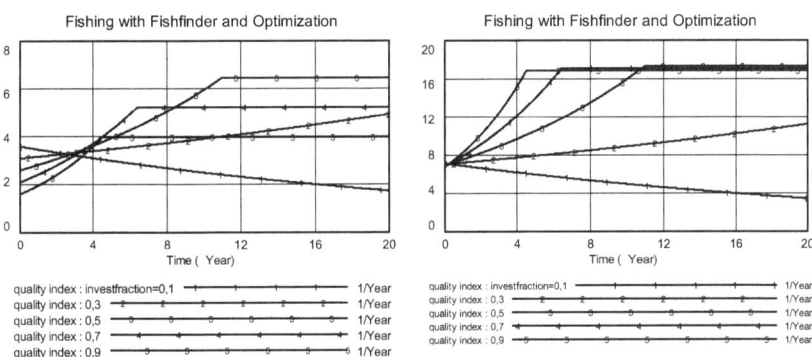

Figure Z410d: If the profit criterion is stressed, an optimal result arises for INVESTMENT FRACTION FOR BOATS = 0.5.
Figure Z410e: If the fish catch volume is important, rapid build-up of the boat fleet is optimal (with INVESTMENT FRACTION FOR BOATS = 0.9).

Exercises

1. Couple module Z115 "State diagram" to model Z410 and generate state trajectory plots for MAX BOAT NUMBER in the range from 25 to 35 (similar to Figure Z410c). Beginning at which MAX BOAT NUMBER does a stable equilibrium point no longer exist? (i.e. the system always collapses). Verify that at this limit for MAX BOAT NUMBER the maximum *profit rate* is also reached. What does this mean in practice?

2. Introduce a variable SPECIFIC GROWTH RATE for *fish* which changes from year to year in the range from 0.5 to 1.5 by random function (small addition to the model). Find a MAX BOAT NUMBER for this case that reliably avoids collapse of the system. Compare *fish catch* and *profit rate* with the results for constant SPECIFIC GROWTH RATE =1 of the default setting.

3. Define as an optimization criterion the profit accumulated over 20 years (supplement the model accordingly). Find (by multiple simulations with different MAX SPECIFIC CATCH VOLUME) a value for this catch parameter which maximizes the profit over 20 years without collapse of the *fish* population, i.e. the parameter should allow sustainable exploitation.

References

Bossel, H. SDS 2004: *Systeme, Dynamik, Simulation – Modellbildung, Analyse und Simulation komplexer Systeme*. Books on Demand, Norderstedt, S. 267-279.

Z411 Tourism

Simulation task

The natural beauty of mountains, lakes, forests, islands, and beaches attracts people. As tourists they want to live in this environment for a couple of weeks and pursue a variety of leisure activities. If the natural environment is stressed only slightly by tourism, any damage will be quickly repaired by natural regeneration of the ecosystem, and the attractiveness of the region remains unaffected. Tourists also spend money in the region. The additional earnings are welcome and will be partially invested to provide still more tourists with a pleasant stay. Many investors want to participate in such a boom. Initially it is hard to imagine how it all might end, as in so many other places: beaches clogged with high-rise hotels; dirty beaches, streams, and lakes; mountain sceneries ruined by concrete construction, highways, and ski lifts. Finally, tourists stay away, leaving the natives in bankruptcy, unemployment, and poverty in a ruined environment.

Is this dynamics unavoidable? Under what conditions can a region be used permanently for tourism without losing its attractiveness? A simulation model can help in trying to describe and understand the dynamics of the system arising from the interaction of tourism with the local environment.

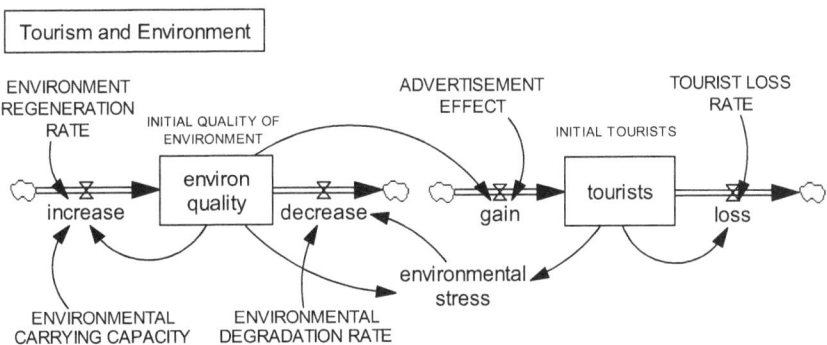

Figure Z411a: Simulation diagram for the tourism-environment system.

Simulation model

Figure Z411a shows the simulation diagram for the simplest possible representation of the dynamic interaction of *environmental quality* and *tourists*. (Somewhat more complex representations are found in models Z412 "Tourism dynamics".) The corresponding model equations are listed in the following. Normalized state variables of order "1" are used since we are primarily interested in determining the possible dynamics.

The *increase* of *environ quality* is determined by ENVIRONMENT REGENERATION RATE and changes as a logistic function limited by ENVIRONMENTAL CARRYING CAPACITY. Without tourism *environ quality* would develop up to this limit. Tourists cause *environmental stress* in proportion to the number of *tourists* and *environ quality* – a type of relationship that is also found in predator-prey systems. The resulting *de-*

crease of *environ quality* depends on the specific ENVIRONMENTAL DEGRADATION RATE.

The *gain* of *tourists* depends on *environ quality*; however, this *gain* can be considerably increased by ADVERTISEMENT EFFECT. The *gain* reflects the attractiveness of the region. The number of *tourists* is reduced by *loss* according to TOURIST LOSS RATE. This *loss* could be expressed, for example, by the loss of overnight stays per year, while a *gain* would be expressed by an increase in overnight stays.

Parameters and initial values
INITIAL QUALITY OF ENVIRONMENT = 1 [quality]
INITIAL TOURISTS = 0.1 [tourist]
ENVIRONMENTAL CARRYING CAPACITY = 1 [quality]
ENVIRONMENT REGENERATION RATE = 1 [1/Year]
ENVIRONMENTAL DEGRADATION RATE = 1 [1/(tourist *Year)]
TOURIST LOSS RATE = 1 [1/Year]
ADVERTISEMENT EFFECT = 5 [tourist /(quality *Year)]

Dynamics
environmental stress = tourists * environ quality [quality *tourist]
increase = ENVIRONMENT REGENERATION RATE * environ quality
       *(1 -environ quality / ENVIRONMENTAL CARRYING CAPACITY) [quality /Year]
decrease = ENVIRONMENTAL DEGRADATION RATE *environmental stress
       [quality /Year]
environ quality = INTEG (+increase -decrease, INITIAL QUALITY OF
       ENVIRONMENT) [quality]
gain = ADVERTISEMENT EFFECT *environ quality [tourist /Year]
loss = TOURIST LOSS RATE *tourists [tourist /Year]
tourists = INTEG (+gain -loss, INITIAL TOURISTS) [tourist]

Simulation time parameters
INITIAL TIME = 0 [Year]
FINAL TIME = 10 [Year]
TIME STEP = 0.02 [Year]

Simulation results

Figure Z411b shows the temporal plots of *environ quality*, *tourists* and *environmental stress* for the default parameter setting. The number of *tourists* increases rapidly at first from a small initial value to almost 2 (normalized quantity!) and then quickly decreases again to an equilibrium value of 0.833. The *environ quality* is quickly reduced from the initial value of 1 to the equilibrium value of 0.167. (The equilibrium state follows from the condition $dz/dt = \mathbf{0}$.) The location of the (stable) equilibrium point is independent of initial conditions.

Figure Z411c shows state trajectories and location of equilibrium points in the state diagram (*environ quality* vs. *tourists*) as a function of ADVERTISEMENT EFFECT. Larger ADVERTISEMENT EFFECT leads to higher number of *tourists* as expected, but also causes stronger permanent reduction of *environ quality*. In particular, more effective advertisement will lead to much stronger initial increase of the number of *tourists*. Correspondingly, the eventual collapse of the number of *tourists* is much stronger. Obviously, the development dynamics of this system (particularly its equilibrium

state) depends very strongly on ADVERTISEMENT EFFECT. The ENVIRONMENTAL DEG-
RADATION RATE from tourism also has a decisive effect on development, as does EN-
VIRONMENT REGENERATION RATE which determines how much tourism the environ-
ment can sustainably support.

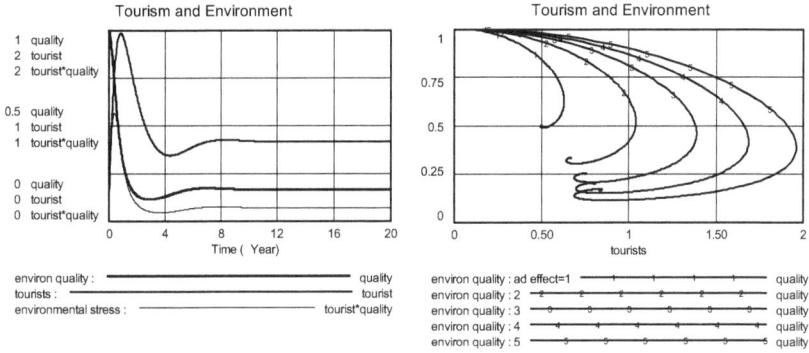

Figure Z411b: Environmental degradation reduces the attractiveness of the region
and causes the initially high number of tourists to drop again.
Figure Z411c: Greater advertising efforts will bring in more tourists in the short run,
but in the long run they cause permanent reduction of the quality of the environment.

Exercises

1. Examine – by simulations with successive change of parameters, and by plotting
the results in a common diagram (similar to Figure Z411c) – the influence of the different parameters on system dynamics and equilibrium state.
2. Determine analytically the equilibrium state for *environ quality* and *tourists* as function of ADVERTISEMENT EFFECT and ENVIRONMENTAL DEGRADATION RATE and plot
results as function of these parameters.
3. Under what circumstances do particularly strong fluctuations of the state variables
appear?
4. How can a gradual development of tourism be achieved while maintaining high
environmental quality

Z412 Tourism dynamics

Simulation task

Systems whose processes follow the laws of physics, like systems in all field of technology, can be uniquely described in terms of mathematical expressions. The physics of a vehicle suspension or of an electronic circuit allows only a single correct mathematical description. The situation is different for dynamic systems whose processes cannot be described with the precision of a physical system – such as ecological and social systems. For example, the system "environment and tourists" presented in model Z411 could certainly also be modeled in other ways. In such cases the question arises whether the chosen model formulation really describes the system correctly, and whether a model builder with different knowledge, other experiences and/or a different cultural background would not perhaps develop an also plausible, but different model formulation that perhaps produces completely different results.

The problem setting of tourism dynamics is well suited for examining this question. The simulation task defined in the following was presented in exactly the same wording to a variety of project teams from quite different educational and cultural backgrounds in several countries of Europe and Southeast Asia. The groups had to develop a simulation model within about three hours without any further explanations being given. Of the numerous models developed at different occasions, four are documented in the following. Although the models differ considerably in their approach, they generate comparable system dynamics leading to the same conclusions. This leads to the expectation that in other cases also, where system complexity does not allow unique mathematical description, dynamic modeling can generate relatively reliable conclusions. This assumption must be carefully tested in every case. However, if models are developed by project teams consisting of people with widely different backgrounds and working independently, and if the models they create lead to comparable conclusions despite differences of formulation, this increases considerably the validity and usefulness of their modeling and simulation results.

Before describing the four models and representative results, the problem description and modeling task handed out to the project groups is reproduced in the following.

Modeling and simulation task: Tourism dynamics of the "Silver bay" region

Problem description
(using direct quotations from New Straits Times, Kuala Lumpur, July 1989)

The Silver Bay island region in the Pacifican Republic has so far been left largely undeveloped. Most of the residents are fishermen living in houses built on stilts close to the sea front. They obtain a meager income from the sale of their catches. Overall, the standard of living in the fishing villages is very low.

The Silver Bay islands are blessed with unspoiled beaches of white sand, unpolluted waters, tropical vegetation, and a pleasant sunny climate: ideal conditions for tourists from industrial countries.

The government of the Pacifican Republic is aware of the enormous development potential of the Silver Bay islands for tourism. It is urging fishermen to discard their traditional ways of thinking and to turn their villages into tourist attractions in order to ensure a higher standard of living for themselves. The government will be providing basic amenities and infrastructure such as roads, electricity, water supply, jetties, and better housing, and will assist in the construction of hotels, chalets, fishing and water sport facilities, and an airport.

The plans are not unopposed, however. Leaders of the opposition party, as well as influential local residents, and scientists of the Pacifican State University, which maintains a world-renown scientific field station on tropical ecology on one of the islands are critical of the plans, maintaining that tourism development would probably eventually destroy the fragile ecology of the islands, thereby also destroying their charm and their tourism attractiveness.

In the face of this political controversy, the government is interested in obtaining an impartial assessment of the possible impacts of the development of tourism in the area. The Ministry of Tourism therefore has decided to ask the well-established consulting firm SysTour to perform a systems study of the long-range dynamic impacts of tourism development in the Silver Bay region.

Your task

You are a systems analyst at SysTour, and are asked to develop a systems model of tourism development in the Silver Bay region together with other specialists. The model should be as simple as possible, but it should be able to describe reliably the major impacts on the region as function of time, under the following alternative assumptions:

A. Everything remains the same; information about the region spreads by word of mouth only, and only a small and constant number of tourists visit the area.

B. Tourism in the area is promoted by a major international advertisement campaign.

Instructions

Phase I: Influence diagram

1. Formulate a verbal model of the relationships important to the system's dynamics.
2. Identify the system quantities necessary for the system description.
3. Use the information from Steps 1 and 2 to draw the corresponding influence diagram (causal loop diagram). (Steps 1 to 3 can be executed at the same time.)
4. Identify (in the diagram) those system variables which are best used to describe the behavior of the system.
5. Give an intuitive assessment of the development of the system for alternatives A and B, and sketch the expected time plot for key variables identified in Step 4.
6. Report your model concept in plenum and explain your expectations concerning the system's dynamics.

Note: If ecosystems remain undisturbed over a long period of time, they develop in the direction of a mature state (climax), i.e. a dynamic equilibrium which corresponds to the ecological carrying capacity of the region.

Phase II: Simulation diagram

7. Identify in your model system: (1) state variables, (2) system parameters, (3) ex-

ogenous quantities, (4) rates of change of state variables, and (5) intermediate variables.

8. Modify the influence diagram by introduction of the symbols for the different types of system quantities (new diagram!)

9. Specify what type of mathematical operation is to be performed to compute a given quantity (integration, addition, multiplication etc.).

10. Formulate the corresponding mathematical relationships (formulae or functions) that allow the computation of each variable from its input variables (and time).

11. Check whether all quantities can be calculated in the resulting simulation diagram.

12. Present the simulation diagram in plenum for discussion.

Note: Since accurate empirical data are not available, and only the qualitative properties of the dynamics are of interest, use relative (normalized) quantities having a reference value of "1" for the initial system state.

Phase III: Simulation

13. Implement the model using a suitable program for computer simulation (including numerical integration of state variables).

14. Quantify the model for the two alternative scenarios A and B.

15. Simulate dynamic development for both cases.

16. Compare the results with your previous intuitive assessment (Step 5).

17. Change and adjust system structure and parameters (within realistic limits only!) until you are relatively certain that the model describes actual dynamics reasonably well.

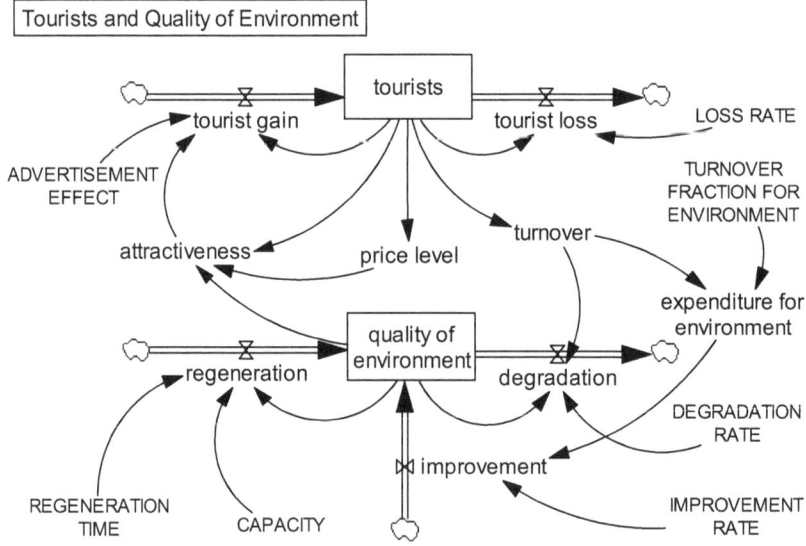

Figure Z412Aa: Simulation diagram for the system "Tourists and quality of the environment".

Simulation model Z412A "Tourists and quality of the environment"

The simulation diagram in Figure Z412Aa and the following model instructions document this model. It contains the two state variables *tourists* and *quality of environment*. These represent variables which are relative and normalized to unity "1". The number of *tourists* grows by *tourist gain* and shrinks by *tourist loss*. *Tourist gain* depends on *attractiveness* and ADVERTISEMENT EFFECT. As the number of *tourists* increases, the *price level* increases. However, these two factors also contribute to a loss of *attractiveness* which otherwise corresponds to the *quality of environment*. Environmental *degradation* is a measure of environmentally damaging impacts of tourism, expressed in financial *turnover* corresponding to number of *tourists*. Part of *turnover* is available as *expenditure for environment* to pay for *improvement* of *quality of environment*. The TURNOVER FRACTION FOR ENVIRONMENT therefore has considerable influence on the dynamics.

Parameters (initial values in INTEG statements)
ADVERTISEMENT EFFECT = 5 [1/Year]
LOSS RATE = 0.5 [1/Year]
TURNOVER FRACTION FOR ENVIRONMENT = 0 [1]
DEGRADATION RATE = 0.1 [1/Year]
IMPROVEMENT RATE = 1 [1/Year]
CAPACITY = 1 [1]
REGENERATION TIME = 10 [Year]

Dynamics
price level = tourists [1]
attractiveness = quality of environment /(tourists *price level) [1]
tourist gain = attractiveness *ADVERTISEMENT EFFECT *tourists [1/Year]
tourist loss = LOSS RATE *tourists [1/Year]
tourists = INTEG (+tourist gain -tourist loss, 1) [1]
turnover = tourists [1]
expenditure for environment = turnover *TURNOVER FRACTION FOR
      ENVIRONMENT /100 [1]
degradation = turnover * quality of environment *DEGRADATION RATE [1/Year]
improvement = expenditure for environment *IMPROVEMENT RATE [1/Year]
regeneration = (quality of environment /REGENERATION TIME) *(1 -quality of
      environment / CAPACITY) [1/Year]
quality of environment = INTEG (regeneration +improvement -degradation, 1) [1]

Simulation time parameters
INITIAL TIME = 0 [Year]
FINAL TIME = 20 [Year]
TIME STEP = 0.05 [Year]

Simulation results for model Z412A "Tourists and quality of the environment"

Figure Z412Ab shows the time plots for *quality of environment*, *tourists* and *degradation* for the default parameter setting. As the number of *tourists* increases initially, the *quality of environment* deteriorates. As a consequence the number of *tourists* goes down because of decreasing *attractiveness* after reaching an initial maximum.

Figure Z412Ac shows the state trajectories (*quality of environment* vs. *tourists*) for different values of ADVERTISEMENT EFFECT. Stronger ADVERTISEMENT EFFECT leads to quicker and stronger increase of *tourists,* however, it also reduces *quality of environment* more strongly and permanently. The system drifts to an equilibrium state dependent on ADVERTISEMENT EFFECT: Stronger advertising leads to higher number of *tourists* but lower *quality of environment.*

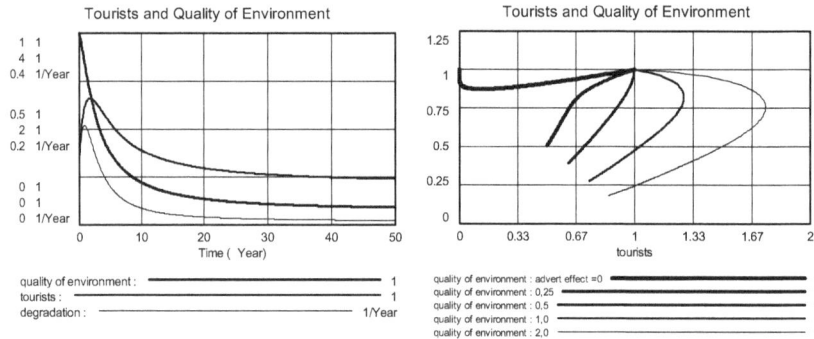

Figure Z412Ab: Quality of the environment and number of tourists decline strongly in the long run.
Figure Z412Ac: More advertisement means more tourists and lower quality of environment. Tourists eventually disappear if there is no advertisement at all.

Simulation model Z412B "Tourist, environment, hotels"

This model is documented in the simulation diagram of Figure Z412Ba and the following model instructions. In this case three state variables *quality of environment, tourists* and *hotels* are used. Here again relative quantities normalized to unity "1" are used since the qualitative dynamics of this system are of primary interest.

The *quality of environment* changes by environmental *degeneration* and *regeneration,* and can be improved by INVESTMENT IN ENVIRONMENT. Environmental *degeneration* is dependent on the number of *tourists.* The number of *tourists* changes by *tourist loss* and *tourist gain.* The latter depends on *attractiveness* which is a function of *quality of environment,* availability of *hotels,* and ADVERTISEMENT. The *quality of environment* (i.e. quality of coastal waters) determines *income from fishery* which together with *income from tourism* determines *net income,* after accounting for COST OF LIVING and INVESTMENT IN ENVIRONMENT. *Savings* are partly used for *private investment* in *hotels.* In addition, *public investment* contributes to *investment* in *hotels* through INVESTMENT SUBSIDY by government. The stock of *hotels* is subject to *depreciation* by gradual decay corresponding to DEPRECIATION RATE.

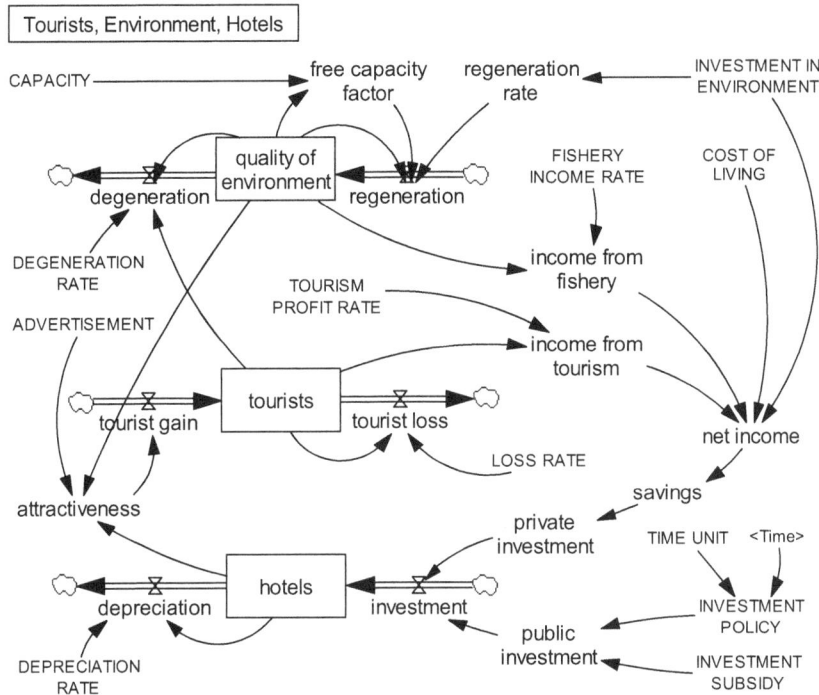

Figure Z412Ba: Simulation diagram for the system "Tourists, environment, hotels".

Parameters and initial values
CAPACITY = 0.05 [1]
DEGENERATION RATE = 1 [1/Year]
INVESTMENT IN ENVIRONMENT = 0.2 [1/Year]
TOURISM PROFIT RATE = 1 [1/Year]
FISHERY INCOME RATE = 1 [1/Year]
COST OF LIVING = 1 [1/Year]
ADVERTISEMENT = 1 [1/Year]
LOSS RATE = 0.3 [1/Year]
DEPRECIATION RATE = 1/20 [1/Year]
INVESTMENT POLICY = WITH LOOKUP (Time / TIME UNIT, ([(0, 0) -(20 ,2)], (0, 1),
     (2, 1), (4, 1), (6, 0), (20, 0))) [1]
INVESTMENT SUBSIDY = 1 [1/Year]
TIME UNIT = 1 [Year]

Dynamics
free capacity factor = CAPACITY -quality of environment [1]
degeneration = DEGENERATION RATE *tourists *quality of environment [1/Year]
regeneration rate = 1 +10 *INVESTMENT IN ENVIRONMENT [1/Year]
regeneration = 0.05 *free capacity factor *quality of environment *regeneration rate
     [1/Year]

quality of environment = INTEG (+regeneration -degeneration, 1) [1]
attractiveness = ADVERTISEMENT *hotels *quality of environment [1/Year]
tourist gain = IF THEN ELSE(attractiveness > 0, attractiveness, 0) [1/Year]
tourist loss = LOSS RATE *tourists [1/Year]
tourists = INTEG (+tourist gain -tourist loss, 0) [1]
income from fishery = quality of environment *FISHERY INCOME RATE [1/Year]
income from tourism = tourists *TOURISM PROFIT RATE [1/Year]
net income = income from fishery +income from tourism -COST OF LIVING
    -INVESTMENT IN ENVIRONMENT [1/Year]
savings = IF THEN ELSE(net income >0, 0.5*net income, 0) [1/Year]
private investment = 0.5 *savings [1/Year]
public investment = INVESTMENT SUBSIDY *INVESTMENT POLICY [1/Year]
investment = 0.8 *private investment +public investment [1/Year]
depreciation = hotels *DEPRECIATION RATE [1/Year]
hotels = INTEG (+investment -depreciation, 0) [1]

Simulation time parameters
INITIAL TIME = 0 [Year]
FINAL TIME = 20 [Year]
TIME STEP = 0.05 [Year]

Simulation results for model Z412B "Tourists, environment, hotels"

The time response of this model for the default parameter values is shown in Figure
Z412Bb. With an increase of *tourists* the *quality of environment* deteriorates quickly.
The number of *hotels* also increases quickly because of the *investment* made possible
by tourism, but attains its maximum only some years after the maximum of *tourists*.
After that the number of *hotels* goes down again gradually. In the long run the system
drifts towards an equilibrium point with low *quality of environment*, low number of
tourists and a reduced number of *hotels*.

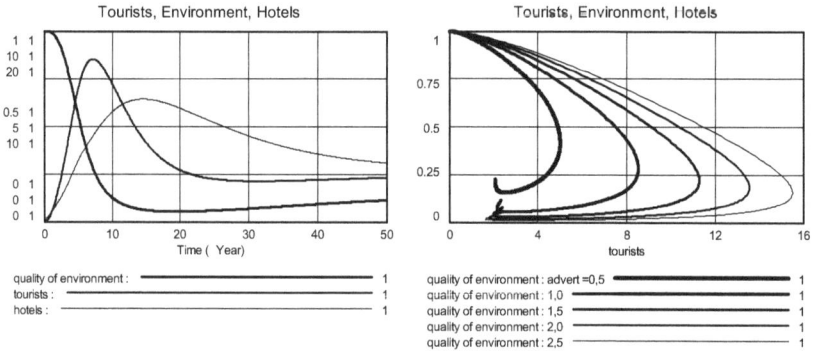

Figure Z412Bb: In the long run an equilibrium state is approached with low quality of
environment and low number of tourists.
Figure Z412Bc: Stronger advertisement leads in the long run to lower quality of environment and almost constant low number of tourists.

The influence of advertisement on the system's dynamics and its equilibrium state is shown in Z412Bc. By a strong ADVERTISEMENT effort the number of *tourists* is initially "hyped" to a multiple of the number that can be sustainably supported in system equilibrium in the long run. This produces corresponding unnecessary *investment* in hotels. In each of the cases shown a similar low equilibrium value develops for *tourists*, while the equilibrium value for *quality of environment* depends strongly on ADVERTISEMENT. More of it produces a lower *quality of environment*.

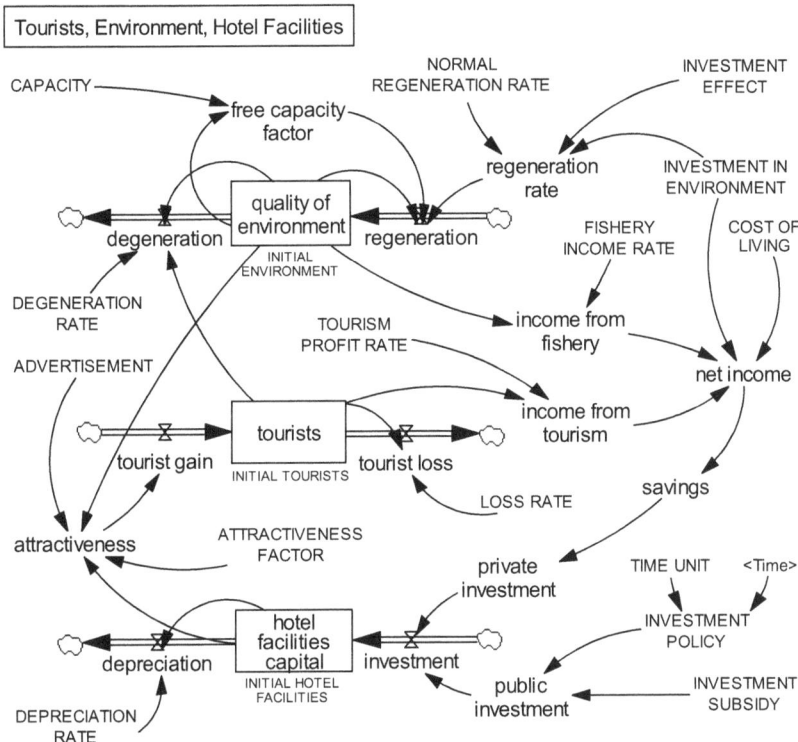

Figure Z412Ca. Simulation diagram for "Tourists, environment, hotel facilities" system

Simulation model Z412C "Tourists, environment, hotel facilities"

Figure Z412Ca shows the simulation diagram for this model. The corresponding model instructions are listed in the following. The model uses the same system structure as model Z412B, but employs a fairly realistic quantification. The *quality of environment* is represented by the fish population (in metric tons of fish) in the waters of the bay. The variable *tourists* is quantified by people per year (more appropriate would be the number of overnight stays per year). The *hotel facilities capital* is measured in terms of their investment value (in $). The different *incomes* are accounted for in $/year.

Parameters and initial values
INITIAL ENVIRONMENT = 1000 [fish t]
INITIAL TOURISTS = 0 [people/Year]
INITIAL HOTEL FACILITIES = 0 [$]
CAPACITY = 1000 [fish t]
DEGENERATION RATE = 0.05 [(1/Year)/(people /Year)]
NORMAL REGENERATION RATE = 0.05 [1/Year]
INVESTMENT IN ENVIRONMENT = 0.2 [1]
TOURISM PROFIT RATE = 1000 [$/((people *Year)/Year)]
FISHERY INCOME RATE = 100 [$/(fish t*Year)]
COST OF LIVING = 100000 [$/Year]
ADVERTISEMENT = 1 [1]
ATTRAKTIVITÄTS FAKTOR = 1e-006 [(people /Year/Year)/($* fish t)]
LOSS RATE = 0.3 [1/Year]
DEPRECIATION RATE = 1/20 [1/Year]
INVESTMENT POLICY = WITH LOOKUP (Time / TIME UNIT, ([(0, 0) -(60, 2)], (0, 1),
        (2, 1), (4, 1), (6, 0), (20, 0), (50, 0))) [1]
INVESTMENT SUBSIDY = 100000 [$/Year]
INVESTMENT EFFECT = 10 [1]
TIME UNIT = 1 [Year]

Dynamics
degeneration = DEGENERATION RATE *tourists *quality of environment *0.001 [fish
        t/Year]
regeneration rate = (1 +INVESTMENT EFFECT *INVESTMENT IN ENVIRONMENT)
        *NORMAL REGENERATION RATE [1/Year]
regeneration = free capacity factor *quality of environment *regeneration rate
        [fish t/Year]
quality of environment = INTEG (regeneration -degeneration, INITIAL
        ENVIRONMENT) [fish t]
free capacity factor = 1 -(quality of environment /CAPACITY) [1]
attractiveness = hotel facilities capital *quality of environment *ADVERTISEMENT
        *ATTRAKTIVITÄTS FAKTOR [people /(Year*Year)]
tourist gain = IF THEN ELSE(attractiveness > 0, attractiveness, 0)
        [people /(Year*Year)]
tourist loss = LOSS RATE *tourists [people /(Year*Year)]
tourists = INTEG (+tourist gain -tourist loss, INITIAL TOURISTS) [people /Year]
income from fishery = quality of environment *FISHERY INCOME RATE [$/Year]
income from tourism = tourists *TOURISM PROFIT RATE [$/Year]
net income = (income from fishery +income from tourism -COST OF LIVING)
        *(1 - INVESTMENT IN ENVIRONMENT) [$/Year]
savings = IF THEN ELSE (net income > 0, 0.5 *net income, 0) [$/Year]
private investment = 0.5 *savings [$/Year]
public investment = INVESTMENT SUBSIDY *INVESTMENT POLICY [$/Year]
investment = private investment +public investment [$/Year]
depreciation = hotel facilities capital *DEPRECIATION RATE [$/Year]
hotel facilities capital = INTEG (+investment -depreciation, INITIAL HOTEL
        FACILITIES) [$]

Simulation time parameters
INITIAL TIME = 0 [Year] , FINAL TIME = 50 [Year]
TIME STEP = 0.05 [Year]

Simulation results for model Z412C "Tourists, environment, hotel facilities"

The time plot of state variables *quality of environment, tourists,* and *hotel facilities capital* for the default parameter setting is shown in Figure Z412Cb. The time development is not qualitatively different from that in model Z412B; however, the results are stretched on the time axis as a result of the different quantification.

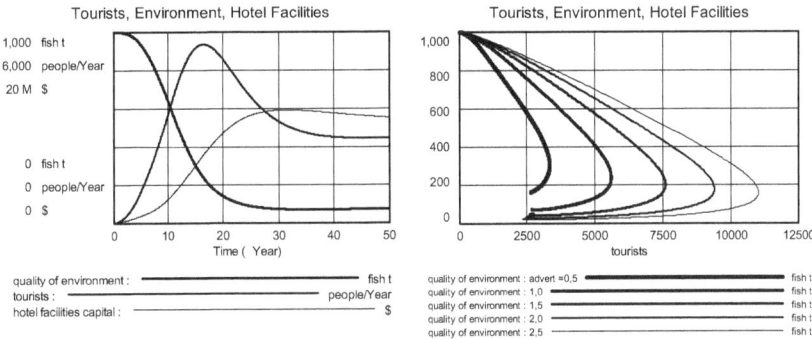

Figure Z412Cb: In the long run the loss of quality of environment is partly compensated for by the attractiveness of hotel facilities.
Figure Z412Cc: Advertisement does not attract more tourists in the long run; however, it leads to a lower quality of environment.

The state diagram for different values of the parameter ADVERTISEMENT (Figure Z412Cc) also shows the same general behavior as in the case of model Z412B. Here also a low equilibrium value develops that is relatively independent of ADVERTISEMENT for *tourists* in the long run, while the equilibrium state of *quality of environment* is reduced substantially more strongly for stronger ADVERTISEMENT.

Comparison of the results for models Z412B and Z412C reveals that model behavior is primarily dependent on system structure, which produces qualitatively equivalent behavior although quantification of both models is significantly different.

This observation means that qualitatively correct conclusions concerning system behavior can be expected even for an uncertain data base where quantification has to rely on rough estimates. Determining the system structure as exactly as possible must always have the greatest priority. Only in rare cases (where for example a switching process depends on an exact balance of countervailing effects) is high data precision required in real-structure models to produce reliable conclusions about behavior.

Simulation model Z412D "Environment, tourists, infrastructure, population"

A somewhat different model concept for the same simulation task is shown in the simulation diagram of Figure Z412Da. The corresponding model instructions are listed in the following. In this case again relative dimensionless state variables were used: *quality of environment, tourists, infrastructure,* and *population.* The state variable *population* was introduced to allow a *population increase* in response to the economic development of the region.

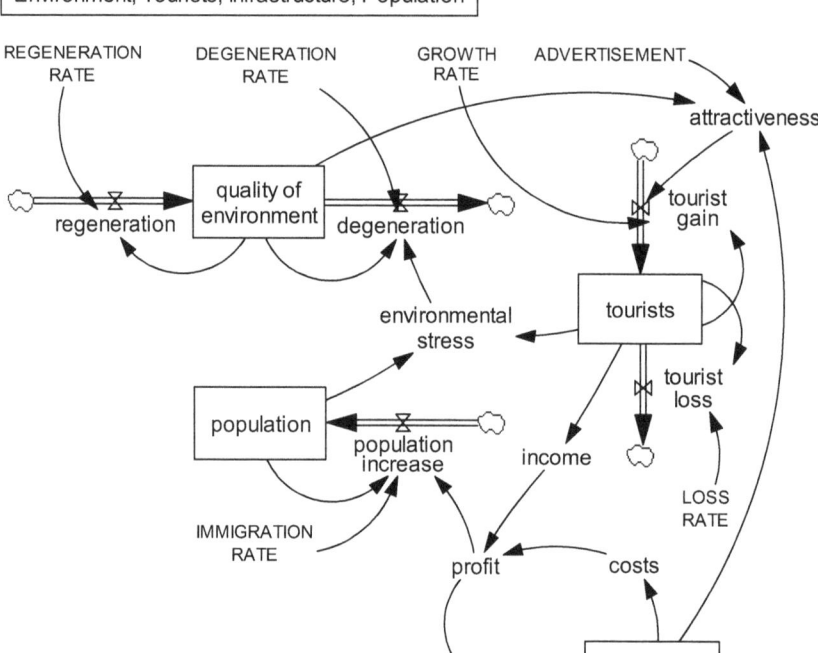

Figure Z412Da: Simulation diagram for the system "Environment, tourists, infrastructure, population".

The *quality of environment* changes by (natural) *regeneration* and *degeneration* of the environment which is due to *environmental stress* caused by *tourists* and *population*. The number of *tourists* diminishes by *tourist loss*, and increases by *tourist gain* which depends on the *attractiveness* of the region. This is a function of the *quality of environment* and the state of the *infrastructure*. *Income* from tourism and the *costs* of infrastructure determine the *profit*, which is used partly for *investment* in infrastructure, and partly attracts immigrants (at IMMIGRATION RATE), thus causing a *population increase*.

Parameters (initial values in INTEG statements)
REGENERATION RATE = 0.05 [1/Year]
DEGENERATION RATE = 0.01 [1/Year]
GROWTH RATE = 0.5 [1/Year]
ADVERTISEMENT = 2 [1]
LOSS RATE = 0.5 [1/Year]
IMMIGRATION RATE = 0.02 [1/Year]
INVESTMENT RATE = 0.1 [1/Year]

Dynamics

regeneration = REGENERATION RATE *quality of environment [1/Year]

environmental stress = population +4*tourists [1]

degeneration = DEGENERATION RATE *environmental stress *quality of environment [1/Year]

quality of environment = INTEG (regeneration -degeneration, 1) [1]

attractiveness = ADVERTISEMENT *infrastructure *quality of environment [1]

tourist gain = attractiveness *GROWTH RATE *tourists [1/Year]

tourist loss = LOSS RATE *tourists [1/Year]

tourists = INTEG (+tourist gain -tourist loss, 1) [1]

population increase = IMMIGRATION RATE *Profit *population [1/Year]

population = INTEG (population increase, 1) [1]

income = tourists [1]

costs = infrastructure [1]

profit = income -costs [1]

investment = INVESTMENT RATE *profit [1/Year]

infrastructure = INTEG (investment, 1) [1]

Simulation time parameters

INITIAL TIME = 0 [Year]

FINAL TIME = 20 [Year]

TIME STEP = 0.05 [Year]

Simulation results for model Z412D "Environment, tourists, infrastructure"

Figure Z412Db shows the temporal development of *quality of environment, tourists, infrastructure*, and *population* for the default parameter setting. Although the model structure is different, the model behavior is qualitatively equivalent to that of the other models: The *quality of environment* diminishes rapidly while the number of *tourists* increases strongly at first. The *tourists* reach their maximum when the *quality of environment* is all but destroyed. After that the *tourists* gradually disappear again while *infrastructure* and *population* still reach a maximum after some delay and decline gradually thereafter.

The state diagram (*quality of environment* vs. *tourists*) in Figure Z412Dc shows the development for different strength of ADVERTISEMENT. As before, stronger ADVERTISEMENT produces more rapid *tourist gain* and a higher maximum number of *tourists*. But in all cases the *quality of environment* and the number of *tourists* eventually collapse.

For certain parameter constellations, however, this system can also exhibit completely different behavior. Figures Z412Dd and Z412De show time plots and state diagrams for the case where parameters different from the default setting were used: REGENERATION RATE = 0.1, ADVERTISEMENT = 1, IMMIGRATION RATE = 0, INVESTMENT RATE = 0.002. For moderate advertisement, cautious investment, no immigration and good natural regeneration *infrastructure* increases slightly while strong undamped oscillations with a period of about three decades develop for *quality of environment* and *tourists*. In this case the system does not collapse.

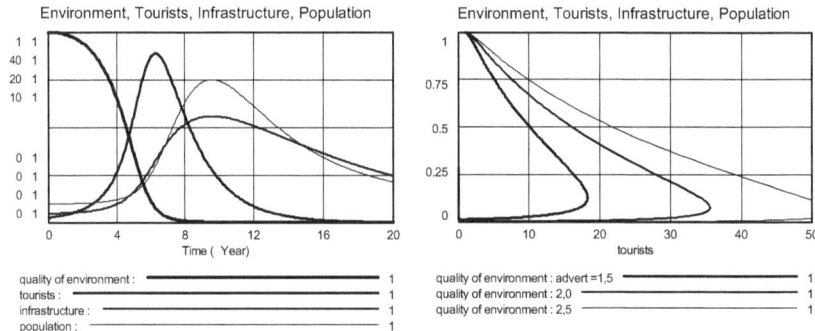

Figure Z412Db: As the environment is destroyed, the number of tourists also declines.
Figure Z412Dc: In this case advertisement cannot prevent the tourist decline.

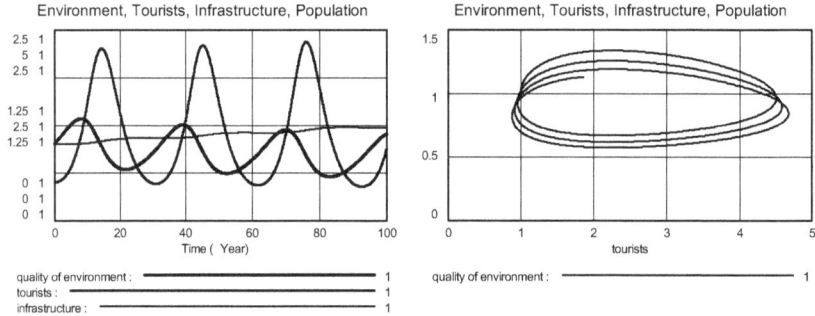

Figure Z412Dd: Cautious expansion of tourism does not lead to collapse.
Figure Z412De: In this case a regular oscillation with a period of about 30 years develops.

Remarks on results with the different simulation models

The four simulation models presented here were independently developed for the same simulation task by different project groups. They employ different structures and quantifications, but exhibit similar behavior. The computed dynamics confirms intuitive assessments and is confirmed by developments in the real world. Modeling and simulation therefore appear to be able to deliver important information concerning the dynamics of processes and system relationships which cannot be precisely defined and must be described in qualitative and fuzzy terms – information that could not be obtained without modeling and simulation.

Exercises

1. Following the steps mentioned above, develop your own model for the tourism problem. As far as possible, develop your own independent concept, and avoid being influenced by the approaches documented here. Can you discover processes or behavioral modes that differ fundamentally from those found above?

2. Examine the behavior of one or several of the models above as function of their parameters (within plausible limits). Identify those parameters that have little influence on the behavior and concentrate further investigations on parameters that strongly influence behavior. Document prominent behavioral modes (oscillations, collapse, equilibrium states) as functions of these parameters. *Note:* The SyntheSim function of the Vensim-PLE simulation software is particularly suitable for examinations of this type.

3. Combine the model instructions of the four models mathematically into two, three, or four differential equations for the state variables. Determine equilibrium points analytically as function of system parameters. Insert the default parameter values and calculate the numerical values of equilibrium states as function of the parameter AD-VERTISEMENT.

4. Expand one of the models so that the annual profit (*income – costs*) can be calculated. Integrate this to obtain total profit (as a function of time). Develop an advertising and investment strategy (using table functions of time, if necessary) which produces maximum total profit over 50 years. Important condition: The strategy must be sustainable even after 50 years, i.e. environment and tourism are not allowed to collapse after this period.

Z413 Forest clearing

Simulation task

In the course of human settlement, formerly wooded areas – like Central Europe or the North American East – have been gradually cleared of their forests to create farmland and produce food for a growing population. As crop yields increased and human populations stabilized, the pressure for clearing forests has essentially ceased in these regions. The "cultivated" landscape of these regions now changes little in its region-specific composition of rural settlements, arable land, and remaining (mostly managed) forests.

In other regions of the world, as in Central America, Laos, and Cambodia, the ruins of vanished civilizations are today overgrown by forest which has reclaimed the land again from which it was cleared by settlers centuries ago. Forest is in many regions of the earth the final stage of natural succession (climax) – an equilibrium state which the ecosystem eventually assumes if it can develop unperturbed by human intervention or natural disasters.

Forests are cleared to harvest their wood or to create areas of arable land. The pressure to clear forests therefore arises from human demand, and it is therefore usually directly connected to population development of a region. However, further processes are also connected to clearing and agriculture: Agriculture takes up nutrients from the soil and opens the land to erosion by wind and water. If the soil is not permanently fertilized and carefully protected from erosion, it loses its fertility, farm yields drop drastically, and the land is finally abandoned and reclaimed by forest in a slow process of ecological succession. Viewed over a span of centuries, a dynamic process has occurred that can be described by a proper simulation model: As a growing population must be fed, more and more forests are cleared. For a while the fertile soil supplies the necessary food, but it gradually loses its fertility. Finally almost all forest has been cleared, but the infertile soil can only support few people any more. People migrate elsewhere, forests gradually develop again on the abandoned land, and fertility returns to the soil as forest litter is decomposed and nutrients recycle. Then the fertile forest soil attracts settlers again, the forest is cleared again for agriculture, and the cycle repeats.

Simulation model

Figure Z413a shows the simulation diagram. The corresponding model instructions are listed in the following. The model uses the state variables *population*, *soil fertility*, *natural forest area* and *farm area*. An interesting feature of this model is the closed cycle of "flows" of land area between *natural forest area* and *farm area*: By *clearing* of *natural forest area* land becomes *farm area* while *farm area* changes again into *natural forest area* by the gradual process of ecological *succession*.

Farm area is used for *food production* according to *soil fertility* and leads to *food supply* which has to meet the *food demand* of the *population*. This demand is determined by PER CAP FOOD DEMAND. The *net growth* of the *population* is determined by *population* level and the NORMAL NET GROWTH RATE (which accounts for births and deaths). The – adequate or inadequate – *food supply* can cause *immigration* (or emigration) corresponding to NORMAL IMMIGRATION RATE. Unrelated to this process a small number of additional IMMIGRANTS is assumed (who resettle the previously abandoned area, for example).

Soil fertility is lost by degradation. The degradation factor increases if the clea-red fraction of land increases. If degradation is low, then regeneration of soil fertility up to its NORMAL PRODUCTIVITY (expressed as per hectare yield in grain cultivation) is possible.

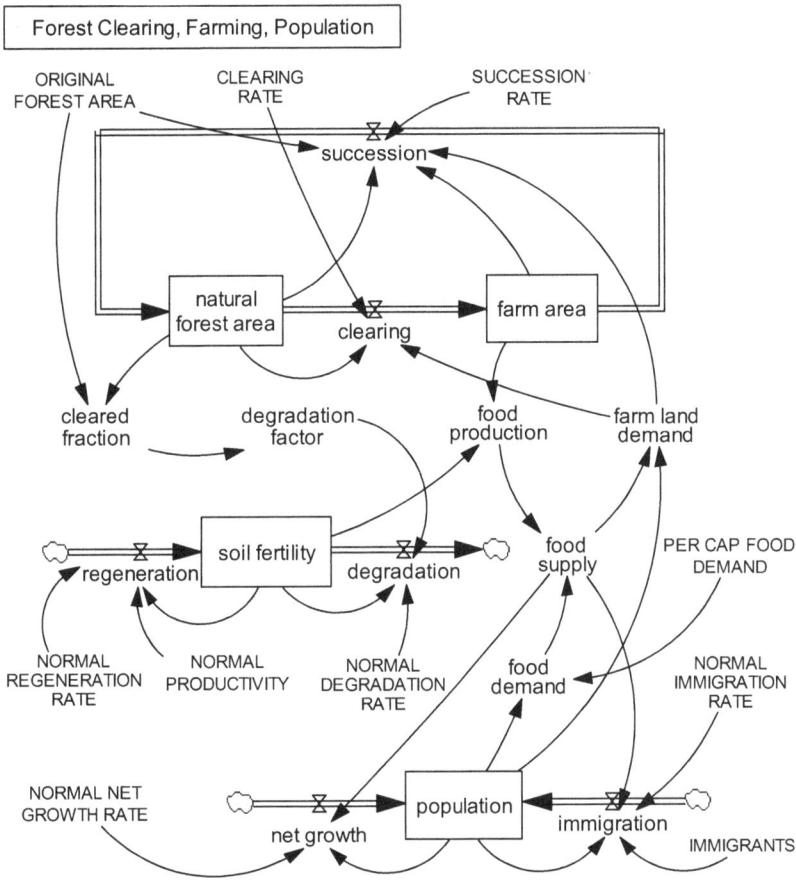

Figure Z413a: Simulation diagram of clearing of forest for farm land.

Parameters and initial values
ORIGINAL FOREST AREA = 10000 [ha]
CLEARING RATE = 0.1 [1/Year]
SUCCESSION RATE = 0.05 [1/Year]
NORMAL REGENERATION RATE = 0.2 [1/Year]
NORMAL PRODUCTIVITY = 2.5 [t/(Year*ha)]
NORMAL DEGRADATION RATE = 0.1 [1/Year]

NORMAL NET GROWTH RATE = 0.025 [1/Year]
IMMIGRANTS = 10 [Mensch/Year]
NORMAL IMMIGRATION RATE = 0.2 [1/Year]
PER CAP FOOD DEMAND = 0.5 [(t/person)/Year]

Dynamics
net growth = NORMAL NET GROWTH RATE *population *food supply [person /Year]
immigration = NORMAL IMMIGRATION RATE *(food supply -1) *population
        +IMMIGRANTS [person /Year]
population = INTEG (+net growth +immigration, 1000) [person]
food demand = PER CAP FOOD DEMAND *population [t/Year]
food supply = IF THEN ELSE (food demand > 0.01, food production /food demand, 0)
        [1]
farm land demand = IF THEN ELSE (population < 1000 :OR: food supply >= 1, 0,
        1 -food supply) [1]
clearing = IF THEN ELSE (farm land demand > 0, farm land demand
        *natural forest area *CLEARING RATE, 0) [ha/Year]
succession = IF THEN ELSE (farm land demand > 0, 0, SUCCESSION RATE
        *agricultural area *((ORIGINAL FOREST AREA -natural forest area)
        /ORIGINAL FOREST AREA)) [ha/Year]
natural forest area = INTEG (+succession -clearing, 9000) [ha]
cleared fraction = (1- (natural forest area /ORIGINAL FOREST AREA)) *100 [1]
agricultural area = INTEG (clearing -succession, 1000) [ha]
regeneration = NORMAL REGENERATION RATE *soil fertility
        *(1-soil fertility/NORMAL PRODUCTIVITY) [t/(Year*Year*ha)]
degradation factor = WITH LOOKUP (cleared fraction, ([(-1, 0) -(100, 5)], (-1, 1), (0, 1),
        (10, 1), (20, 1.03), (30, 1.2), (40, 1.4), (50, 1.65), (60, 1.95), (70, 2.18), (80, 2.48),
        (90, 2.7), (100, 3))) [1]
degradation = soil fertility*degradation factor *NORMAL DEGRADATION RATE
        [t/(Year*Year*ha)]
soil fertility= INTEG (+regeneration -degradation, NORMAL PRODUCTIVITY)
        [(t/ha) /Year]
food production = soil fertility*agricultural area [t/Year]

Simulation time parameters
INITIAL TIME = 0 [Year]
FINAL TIME = 2000 [Year]
TIME STEP = 0.5 [Year]

Simulation results

Figure Z413b shows the temporal development of *population, soil fertility, natural forest area* and *farm area* for the default parameter setting. A cycle with a period of about 200 years develops. At first *natural forest area* is increasingly cleared. In this process a growing *population* secures more *farm area* for its *food production.* Nutrient uptake and erosion cause *degradation* of *soil fertility.* Diminishing crop yields can only feed a decreasing *population. Farm land* is abandoned and left to natural *succession,* and *natural forest area* and *soil fertility* eventually return as *cleared fraction* of land decreases. The region once again becomes promising for cultivation, settlers return, clear the forest and convert it to farm land, and the cycle repeats.

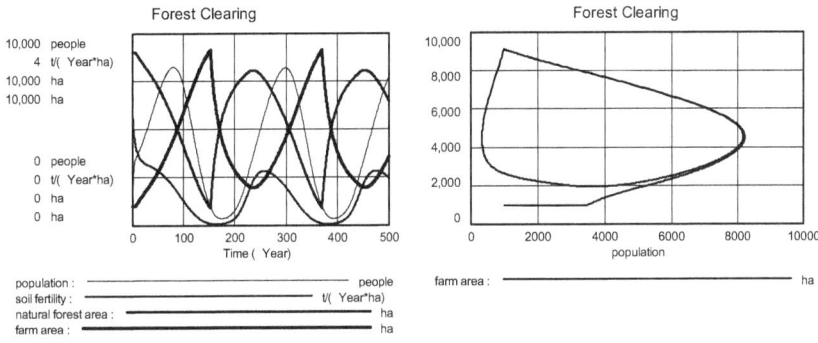

Figure Z413b: The model shows cyclical behavior: clearing of forests and expansion of farm area to feed a growing population is followed by loss of soil fertility and emigration. Subsequently, forest area and soil fertility recover during the natural succession process.

Figure Z413c: In the state diagram, the system state moves on a constantly repeating limit cycle.

Figure Z413c shows a limit cycle appearing in the state diagram (*farm area* vs. *population*) which exactly repeats approximately every two centuries.

Exercises

1. Investigate the influence of the different parameters on the dynamics (primarily the cyclical oscillations) of the model. (Note: Use the SyntheSim feature in VensimPLE.)

2. Document and comment some prominent behaviors and their parameter constellations.

3. Under which circumstances (for which parameters) do steady, noncyclical developments arise?

4. Use a similar approach to model slash-and-burn cultivation of fields, where small clearings are cut in mature natural forest and farmed for about three to five years, are then abandoned to renewed forest growth for about 20 years, after which they are cleared again and used for a few years for field crop cultivation.

Z414 Resource discovery

Simulation task

Today's technology in industrial countries is dependent on permanent exploitation and consumption of nonrenewable fossil and mineral resources. If technology remains unchanged, annual consumption of many resources will increase further for two reasons: (1) increasing industrialization of previously nonindustrialized regions leads to increasing per capita consumption of nonrenewable resources in these regions; and (2) population growth leads to further consumption growth even if per capita consumption should remain the same.

Material resources do not disappear with their use, but they are usually diluted and scattered in the environment by processing, wear, and scrapping to a degree that makes 100 per cent recycling impossible even with great effort. And, of course, energy cannot be recycled at all after use.

The inevitable exhaustion of most exploitable resources is meanwhile more and more obvious; it will lead to supply problems for many resources within the next decades. It can easily be shown that if consumption continues to grow, duplication or even tenfold multiplication of recoverable resources has only insignificant influence on stretching their "life time". To provide a minimum of material prosperity to a growing human population in future, it is mandatory to use materials much more efficiently, i.e. to reduce resource use per unit of material service by change of current technologies. Several approaches can be used alone or in combination: recycling of material, material-saving design, replacement of scarce resources by more widely available resources, and in particular: use of renewable resources, long life cycle of products, design for easy repair, exchange, and overhaul, remanufacturing of machinery etc. Many still largely unused possibilities still exist in this field.

The exploitation dynamics of a resource arises from the interplay between demand and still available supply. At the beginning of the exploitation cycle the available stocks and their discovery rates are very large in proportion to exploitation rates. Exploitation can therefore accelerate exponentially. Gradually the success rate of exploration decreases, however. As (currently known) reserves are depleted by exploitation and by decreasing exploration success, the growth rate of exploitation becomes negative (inflection point). Thereafter, the exploitation rate gradually reaches a maximum and then drops to zero with (exponential) exhaustion of reserves. From the inflection point of the exploitation curve it is therefore possible to draw conclusions concerning the total stock and its life time even without all stocks having been discovered already (Hubbert 1969).

Simulation model

Figure Z414a shows the simulation diagram which is quantified using the model instructions listed in the following. The two state variables are the stock of *discovered resources* and the amount of already *consumed resources*.

The *discovery* of resources follows a logistic function which is defined by MAX DISCOVERY RATE and MAX DISCOVERABLE RESOURCES, and is also enhanced by *discovery effort*. This effort increases as resources are increasingly depleted and become scarcer, i.e. as the amount of *available resources* decreases with respect to the amount of MAX DISCOVERABLE RESOURCES. *Consumption* is proportional to *available re-sources*. It goes down for lower MAX CONSUMPTION RATE and higher RECYCLED RE-SOURCE SHARE.

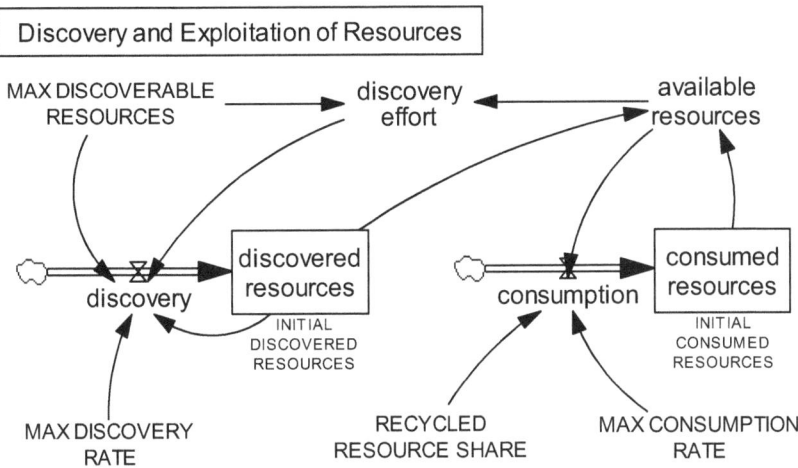

Figure Z414a: Simulation diagram for discovery and exploitation of resources.

Parameters and initial values
INITIAL DISCOVERED RESOURCES = 0.01 [Mio t]
INITIAL CONSUMED RESOURCES = 0 [Mio t]
MAX DISCOVERABLE RESOURCES = 1 [Mio t]
MAX DISCOVERY RATE = 0.1 [1/Year]
MAX CONSUMPTION RATE = 0.1 [1/Year]
RECYCLED RESOURCE SHARE = 0 [1]

Dynamics
discovery effort = 1 -(available resources /MAX DISCOVERABLE RESOURCES) [1]
discovery = MAX DISCOVERY RATE *discovered resources *(1 -(discovered
     resources /MAX DISCOVERABLE RESOURCES)) *discovery effort [Mio t/Year]
discovered resources = INTEG (+discovery, INITIAL DISCOVERED RESOURCES)
     [Mio t]
consumption = (1 -RECYCLED RESOURCE SHARE) *MAX CONSUMPTION RATE
     *available resources [Mio t/Year]
consumed resources = INTEG (+consumption, INITIAL CONSUMED RESOURCES)
     [Mio t]
available resources = discovered resources -consumed resources [Mio t]

Simulation time parameters
INITIAL TIME = 0 [Year]
FINAL TIME = 200 [Year]
TIME STEP = 0.125 [Year]

Simulation results

Figure Z414b shows time plots for the default parameter setting. It is assumed here that annual MAX DISCOVERY RATE corresponds to 1/10 of MAX DISCOVERABLE RESOURCES, and that MAX CONSUMPTION RATE also amounts to 1/10 of this. This leads to a *consumption* maximum after 55 years, and an exhaustion of resources after about 120 years.

The amount of *discovered resources* rises logistically up to its limit MAX DISCOVERABLE RESOURCES. With increasing *available resources* the *consumption* of resources also increases. This leads to an increase in *consumed resources* which is also logistic but is delayed compared to *discovery* of resources. The rate of *discovery* increases up to a maximum, but then goes down to zero with increasing exhaustion of resources. Although delayed, *consumption* of resources shows a similar course. In the course of the development a point is reached where *consumption* exceeds *discovery*. From this point on *available resources* can only decrease.

Decisive for long-term resource availability are the rate of *consumption* and also the RECYCLED RESOURCE SHARE. The influence of the total amount of MAX DISCOVERABLE RESOURCES on resource availability and time period until exhaustion is smaller than expected, since a larger resource supply will also stimulate higher *consumption*.

Figure Z414c shows the time plots for available resources for different RECYCLED RESOURCE SHARE in the range from $r = 0.2$ to 1.0. The latter course is unattainable for physical reasons. All other options (with less than complete recycling) lead to eventual and inevitable exhaustion of the resource even for very high recycling shares.

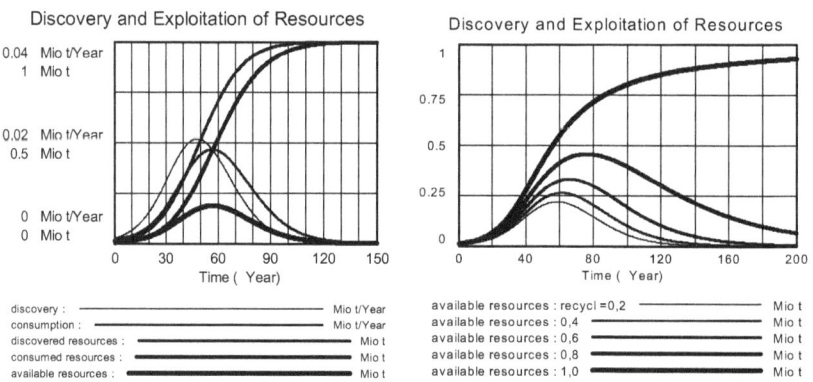

Figure Z414b: The history of resource discovery can be used to estimate the total amount of resources and time to depletion.
Figure Z414c: The more resources are recycled, the more the time to depletion can be shifted into the future.

Exercises

1. By repeated simulation runs, plotting results in a common diagram, examine the influence of parameters MAX CONSUMPTION RATE and RECYCLED RESOURCE SHARE on the time of maximum *consumption* and the point of time where 90 percent of MAX DISCOVERABLE RESOURCES have been used up.

2. Investigate, for an intermediate MAX CONSUMPTION RATE, by what factor the "lifetime" of a resource can be prolonged (without recycling), if the amount of MAX DISCOVERABLE RESOURCES is increased by a factor of 10 and 100.

3. Explain how it is possible to estimate the (unknown) amount of resources and their time of depletion by using the time graph of resource discovery (following Hubbert 1969).

4. Obtain current data for the historical development of discovery and exploitation (= consumption) of mineral oil and natural gas (from the literature or the internet). Use these to quantify the model such that simulation results agree reasonably well with historical data. Simulate future development up to the year 2100.

5. Draw conclusions concerning the future availability and "life time" of mineral oil and natural gas from insights gained from simulations with the model.

References

Bossel, H. 1985: *Umweltdynamik – 30 Programme für kybernetische Umwelterfahrungen*. TeWi, München, S. 361-377.

Bossel, H. 1994: *Umweltwissen – Daten, Fakten, Zusammenhänge*. Springer, Berlin /Heidelberg /New York, S. 101-118.

Hubbert, M. K. 1969: Energy resources. In: *Resources and Man*, National Academy of Sciences – National Research Council. W. H. Freeman, San Franciso.

Z415 Resource extraction and recycling

Simulation task

Smaller initial stock, bigger demand, and lower recycling share will accelerate the depletion of a nonrenewable resource.

For some important resources the almost complete exhaustion of stocks is imminent within the next few decades. Estimates of the likely "life time" of resources must not be based on current consumption rate (providing a "static" life time index) but must account for the probable increase of consumption rate. This leads to the calculation of "dynamic life time", which can be substantially shorter than the static life time.

Calculation of static and dynamic life time can at best serve to determine the bounds of actual life time of a resource. As a resource becomes scarce, its consumption must approach zero – thus lengthening the calculated life time. The relative amount of remaining resources, i.e. scarcity, will therefore determine the development of the consumption rate. If material is recycled, it is important to know how quickly a product is scrapped and material is returned to the production process. A model describing the dynamics of nonrenewable resource use must account for these processes.

Simulation model

Figure Z415a shows the simulation diagram. The corresponding model instructions are listed in the following. The three state variables are: *cumulative resource consumption,* the amount of *resource in use* and the *resource demand.* The units used are [Mt] (millions metric tons) or [Mt/year]; this allows the use of data from resource statistics for carrying out corresponding calculations.

The *growth of resource demand* is computed from INITIAL RESOURCE DEMAND, CONSUMPTION INCREASE TODAY, and MAX DEMAND FACTOR. The (future) development of *resource demand* follows by integration. Together with *supply index* (as an indicator of scarcity) and available amount of *recycled resource* this *resource demand* determines the *resource exploitation rate.* By integration of this rate the *cumulative resource consumption* is determined. Resources corresponding to (annual) rate of *resource exploitation* and *recycled resource* determine the *production rate.* It leads (by integration) to increase of *resource in use,* while the *scrapping rate* corresponding to the length of the PRODUCT LIFE CYCLE takes material out of the *resource in use* stock. A switch RECYCLING 0 or 1 is provided to turn recycling "on" or "off". The influence of different estimates for resource availability can be investigated by adjusting the parameter ORIGINAL RESOURCE RESERVES.

Parameters and initial values
INITIAL RESOURCE DEMAND = 5 [Mt/Year]
INITIAL CUMULATED CONSUMPTION = 0 [Mt]
CONSUMPTION INCREASE TODAY = 2 [1/Year] *in percent per year*
ORIGINAL RESOURCE RESERVES = 100 [Mt]
MAX DEMAND FACTOR = 2 [1]
PRODUCT LIFE CYCLE = 10 [Year]
RECYCLING 0 or 1 = 1 [1]

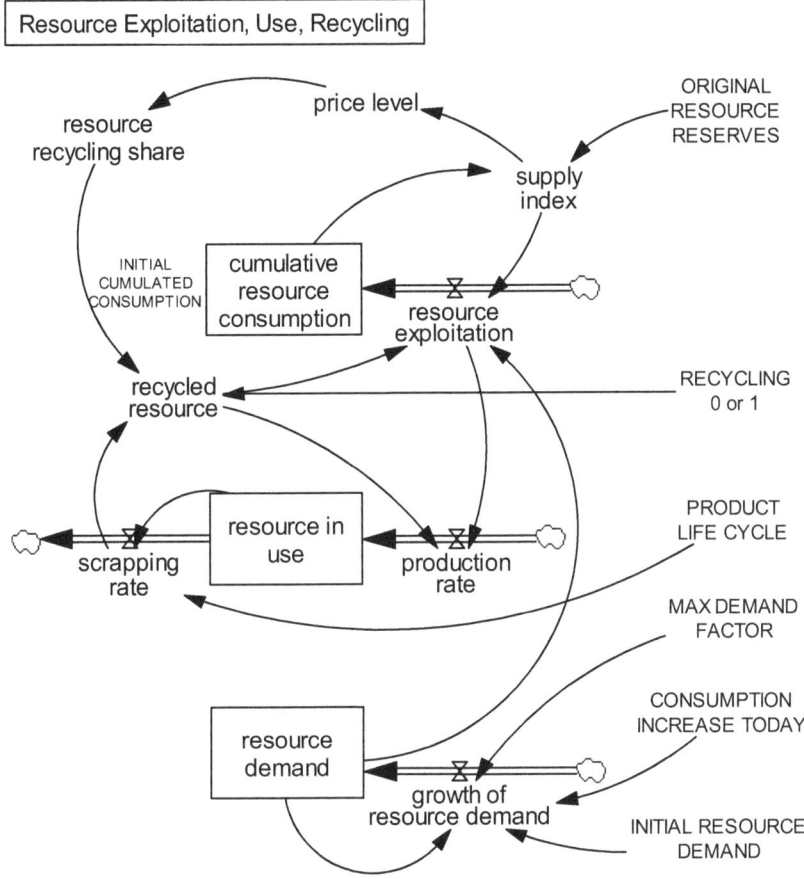

Figure Z415a: Simulation diagram for resource use with recycling.

Dynamics
supply index = (ORIGINAL RESOURCE RESERVES -cumulative resource
    consumption) /ORIGINAL RESOURCE RESERVES [1]
price level = IF THEN ELSE (supply index > 0.1, 1/supply index, 10) [1]
resource recycling share = WITH LOOKUP (price level, ([(0, 0) -(20, 1)], (0, 0), (1, 0),
    (2, 0.2), (5, 0.6), (10, 0.85), (15, 0.95), (20, 0.95))) [1]
recycled resource = scrapping rate *resource recycling share *RECYCLING 0 or 1
    [Mt/Year]
resource exploitation = (resource demand -recycled resource) *supply index [Mt/Year]
cumulative resource consumption = INTEG (+resource exploitation, INITIAL
    CUMULATED CONSUMPTION) [Mt]
production rate = resource exploitation +recycled resource [Mt/Year]
scrapping rate = resource in use /PRODUCT LIFE CYCLE [Mt/Year]
resource in use = INTEG (+production rate -scrapping rate, PRODUCT LIFE CYCLE
    *INITIAL RESOURCE DEMAND) [Mt]

growth of resource demand = (CONSUMPTION INCREASE TODAY /100)
    *resource demand *(1 -resource demand /(INITIAL RESOURCE DEMAND *MAX
    DEMAND FACTOR)) [Mt/(Year*Year)]
resource demand = INTEG (+growth of resource demand, INITIAL RESOURCE DE-
    MAND) [Mt/Year]

Simulation time parameters
INITIAL TIME = 2000 [Year]
FINAL TIME = 2125 [Year]
TIME STEP = 0.25 [Year]

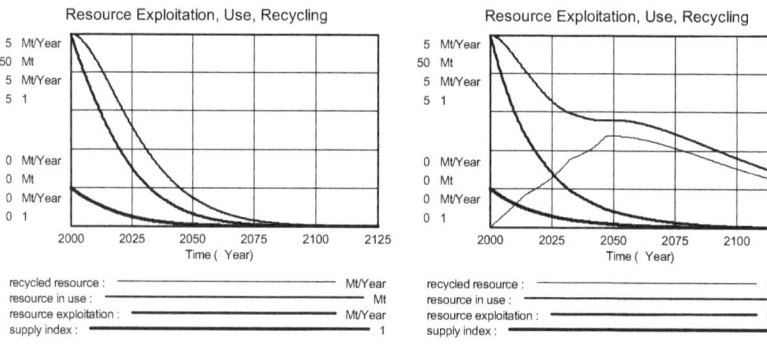

Figure Z415b: Resource availability without recycling.
Figure Z415c: Resource availability with increasing recycling.

Simulation results

Figures Z415b and c show the temporal development of *resource in use, resource exploitation* and *supply index*. In the first case no recycling is scheduled; in the second case it changes according to table function *resource recycling share* as function of price level to 95 percent recycling (corresponding to twentyfold increase of price level).

With the default parameter setting the *resource in use* decreases rapidly if there is no recycling. If resource scarcity (i.e. *supply index*) enforces an increasing share of recycled resource, the amount of *resource in use* can remain much longer at a high level despite strongly decreasing *resource exploitation*. The eventual depletion of the resource cannot be prevented, of course.

Exercises

1. Obtain estimates for current reserves of important nonrenewable resources, for current annual consumption, and for current annual increase of consumption. Calculate static and dynamic life times using the following formulae:

static life time T_s = (estimated reserves)/(current consumption per year)

dynamic life time T_d = $(1/r) \ln[(R \cdot r/C_0)+1]$

where r = annual growth rate of consumption, C_0 = initial annual consumption, R = reserve estimate, T = life time of reserve, ln = natural logarithm. Simulate the consumption dynamics for these resources using different recycling scenarios. Discuss the results in comparison with the calculated static and dynamic life times of the resources. How big are the discrepancies? How reliable are the conclusions? What else should be taken into account in such estimates?

2. Using the model (perhaps after some modification), investigate systematically how the "life time" of a given resource could best be stretched: by discovery of further reserves? by high recycling share? by reduction of demand (e.g. by long-lasting products)? What could each strategy contribute? Make suggestions for a national or global resource strategy. What focus should future research and development have with respect to resources?

References

Bossel, H. 1994: *Umweltwissen – Daten, Fakten, Zusammenhänge*. Springer, Berlin /Heidelberg /New York, S. 101-118.

Council on Environmental Quality: *Global 2000 – Der Bericht an den Präsidenten*. Zweitausendeins, Frankfurt/M. 1980, bes. S. 459-492, 791-810.

Ehrlich, P. R., Ehrlich, A.H., Holdren, J.P., 1977: *Ecoscience – Population, Resources, Environment*. Freeman, San Francisco, bes. S. 391-513, S. 515-531.

Z416 Overshoot and collapse

Simulation task

Model Z405 "Ecosystem collapse" was developed to represent the historical collapse of a deer population after a population explosion following the extermination of predators. Processes of this type are frequent in many areas – not only in ecosystems. This typical system behavior is quite common wherever renewable resources are overused: overfishing, overgrazing, deforestation, firewood crisis. The same dynamics, based on similar structural relationships, also operates in the "world models" which describe the global development of human population, resource use, and environment (cf. models Z605, Z610, Z612 in Bossel 2007 *System Zoo 3*).

The core of such systems consists of a population which depends on a renewable resource. If overused, the regeneration ability of the resource degenerates until a recovery is no longer possible and the resource collapses along with the population that depends on it. The process is reduced to its essential components and processes in the following to examine dynamic behavior and development options.

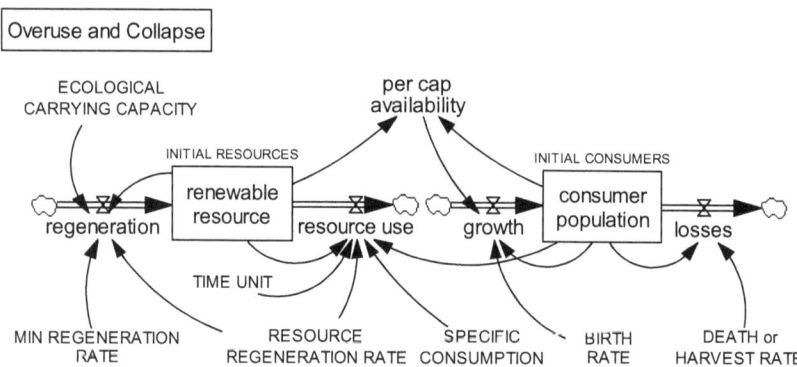

Figure Z416a: Simulation diagram for use of a renewable resource.

Simulation model

The simulation diagram is shown in Figure Z416a. The respective model equations are listed in the following. The two state variables *renewable resource* and *consumer population* are coupled to each other by *resource use* and by the *growth* of the *consumer population* that is made possible by *resource use*.

The *growth* of the *consumer population* is proportional to its size and the *per cap availability* of the *renewable resource*. Since per capita (food) consumption is limited, a Michaelis-Menten saturation is introduced (cf. model Z111 "Density dependent growth" in *System Zoo 1*). The *resource use* (e.g. food consumption) is proportional to *consumer population* and SPECIFIC CONSUMPTION as long as a sufficient amount of *renewable resource* is available. If this is not the case, *resource use* is restricted to the available amount. In the case of malnutrition of the *consumer population*, its rate of *growth* by BIRTH RATE drops below the rate of *losses* by DEATH OR HARVEST RATE.

Logistic development corresponding to ECOLOGICAL CARRYING CAPACITY applies to *regeneration* of *renewable resource*. It regenerates only very slowly if the *renewable resource* stock is small (*regeneration* proportional to the square of *renewable resource* stock), but always with a minimum value of (MIN REGENERATION RATE * ECOLOGICAL CARRYING CAPACITY).

Parameters and initial states
INITIAL CONSUMERS = 0.1 [consumer]
INITIAL RESOURCES = 0.5 [resource]
ECOLOGICAL CARRYING CAPACITY = 1 [resource]
RESOURCE REGENERATION RATE = 1 [1/Year]
MIN REGENERATION RATE = 0.01 [1/Year]
BIRTH RATE = 0.7 [1/Year]
DEATH or HARVEST RATE = 0.5 [1/Year]
SPECIFIC CONSUMPTION = 1 [resource /(consumer*Year)]

Dynamics
regeneration = RESOURCE REGENERATION RATE *renewable resource
    *(renewable resource /ECOLOGICAL CARRYING CAPACITY) *(1 -renewable
    resource /ECOLOGICAL CARRYING CAPACITY) +MIN REGENERATION
    RATE *ECOLOGICAL CARRYING CAPACITY [resource/Year]
resource use = IF THEN ELSE (renewable resource < SPECIFIC CONSUMPTION
    *consumer population, (renewable resource *RESOURCE REGENERATION
    RATE), SPECIFIC CONSUMPTION *consumer population) [resource/Year]
renewable resource = INTEG (+regeneration -resource use, INITIAL RESOURCES)
    [resource]
per cap availability = IF THEN ELSE (consumer population <= 0, 0, renewable
    resource /consumer population) [resource/consumer]
growth = BIRTH RATE *consumer population *(per cap availability
    /(per cap availability +1)) [consumer/Year]
losses = DEATH or HARVEST RATE * consumer population [consumer/Year]
consumer population = INTEG (+growth -losses, INITIAL CONSUMERS) [consumer]

Simulation time parameters
INITIAL TIME = 0 [Year]
FINAL TIME = 200 [Year]
TIME STEP = 0.02 [Year]

Simulation results

Figure Z416b shows the time response for the default parameter setting. These parameters correspond approximately to those of grazing animals on pastureland; the state variables are relative (normalized) quantities. A collapse occurs after about 25 years. After that there is no regeneration; *renewable resource* and *consumer population* remain at very low level. If somewhat smaller BIRTH RATE is selected (0.6 instead of 0.7), ecosystem and grazing animal population remain on a relatively high equilibrium level indefinitely, without collapse.

Starting out from a small initial value of *consumer population*, it initially grows rapidly, resulting in gradual reduction of *renewable resource*. As the resource base erodes, its *regeneration* ability and amount of *renewable resource* diminish, reducing *growth*. After reaching a maximum, *consumer population* collapses with some time delay compared to rapidly collapsing *renewable resource* stock. Because of permanent MIN REGENERATION RATE further existence of the renewable resource at very low level is nevertheless possible.

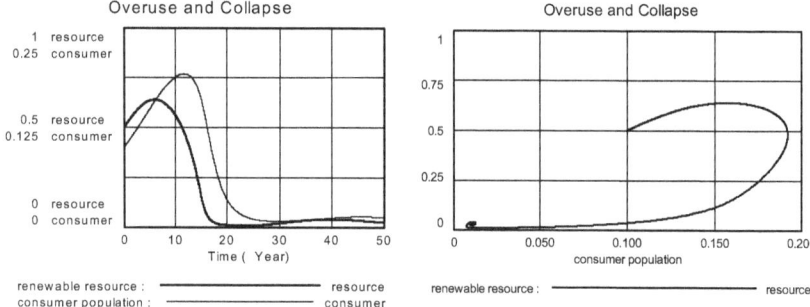

Figure Z416b: Collapse of an ecosystem caused by over-grazing.
Figure Z416c: State trajectory of the collapse with subsequent stabilization at low level.

The state diagram of this simulation is presented in Figure Z416c. Starting out from its initial state at first *consumer population* almost doubles before it then collapses following collapse of *renewable resource*. The state trajectory is captured by an equilibrium point at very low levels of the state variables. The state variables oscillate strongly damped around this point before they come to rest there.

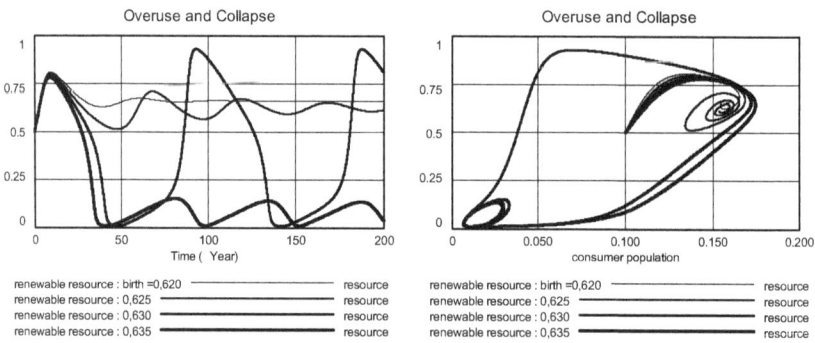

Figure Z416d: Rapid change of behavior resulting from increasing BIRTH RATE: equilibrium of renewable resource at high level is followed by oscillations, then a limit cycle, finally by oscillations at low level of the state variables.
Figure Z416e: State trajectories of the same process in the state diagram.

If one examines the model behavior as function of the parameter BIRTH RATE, then an interesting and rapid change of behavior is observed in a very narrow range of this parameter. Figures Z416d and e show time response and state trajectories for

BIRTH RATE = 0.625, 0.630, and 0.635 (all other parameters are those of the default setting). While for BIRTH RATE = 0.620 an equilibrium state establishes itself at high level of *renewable resource*, for BIRTH RATE = 0.625 a strong, but still damped oscillation around an equilibrium state develops, at high levels of both state variables. If BIRTH RATE is increased slightly (b = 0.630), the damped oscillation develops into a steadily recurring limit cycle which repeats with large amplitude between the previous high and very low state values. If BIRTH RATE is increased further (b = 0.635), then a limit cycle of small amplitude appears around an equilibrium point with low state values. Further increase produces a damped oscillation quickly ending at low final state (cf. Figure Z416c *for b* = 0.7).

Exercises

1. Determine (using the default parameter settings) what hunting quota (= DEATH OR HARVEST RATE) should be chosen for a population of wild animals to stabilize both state variables (*renewable resource* and *consumer population*) at high level.
2. What would have to be done to regenerate a *renewable resource* after collapse?
3. Examine the transition zone indicated by the limit cycle in more detail, also for other plausible parameter combinations.

References

cf. model Z405 "Ecosystem collapse".

Z417 Tragedy of the commons

Simulation task

Commons – that is land owned and used by all inhabitants of a village or a region, in particular pastures. If the number of cattle grazing the commons is restricted to maintain the ability of the renewable resource to regenerate, sustainable (permanent) use of the commons is possible. This particular concept of common ownership and utilization can also be applied to other natural resources: to the fishing grounds of the oceans outside of territorial waters, to fuelwood collection in communal forests, to the use of ground water for irrigation. If this type of use is not strictly regulated – by laws or taboos – it can easily lead to collapse of the renewable resource: The possibility of using a renewable resource in common ownership for individual advantage can tempt individuals to increase their personal profit by investment in additional "means of production" (such as additional heads of cattle, a larger fishing vessel, fish locator technology, another irrigation well). If strict limitation does not exist, this process will cause overuse leading finally to collapse of the resource base. But where must limits be drawn? Under what conditions can a resource in common ownership be used sustainably and profitably by all?

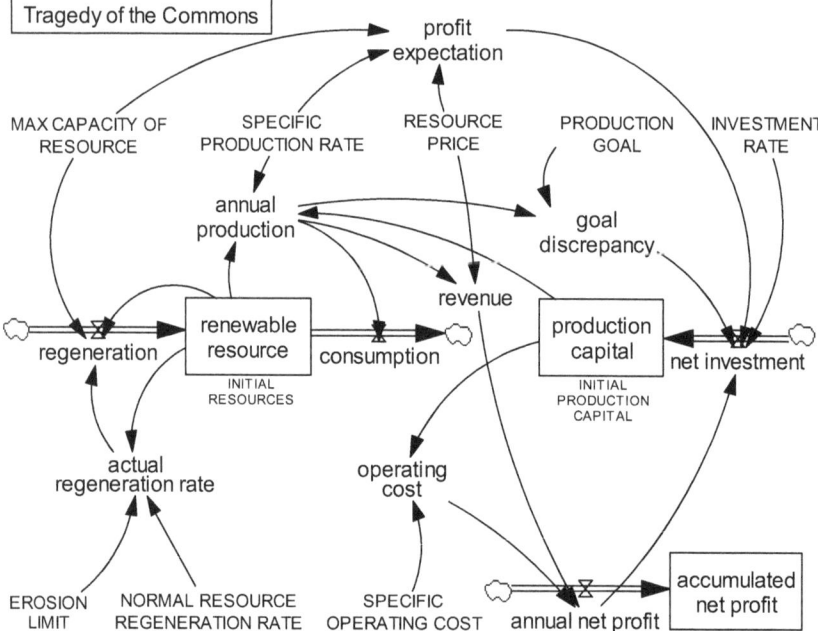

Figure Z417a: Simulation diagram for the tragedy of the commons.

Simulation model

Figure Z417a shows the simulation diagram for this system. The model instructions are listed in the following.

The *renewable resource* has a logistic growth limitation with MAX CAPACITY OF RESOURCE and an initially exponential *regeneration* with specific NORMAL RESOURCE REGENERATION RATE. *Consumption* of the *renewable resource* corresponds to *annual production*. This is proportional to the stock of *production capital* for use of the resource (cattle, equipment, machines) as well as to SPECIF PRODUCTION RATE of these means of production, and the current amount of *renewable resource*. *Net investment* in *production capital* is proportional to the *goal discrepancy* existing with respect to PRODUCTION GOAL and *annual net profit*. For smaller *profit expectation* more *net investment* is made to make up for losses.

Parameters and initial values
INITIAL PRODUCTION CAPITAL = 0.01 [$]
INITIAL RESOURCES = 1 [t]
MAX CAPACITY OF RESOURCE = 1 [t]
NORMAL RESOURCE REGENERATION RATE = 0.1 [1/Year]
EROSION LIMIT = 0.05 [t]
INVESTMENT RATE = 0.1 [1/Year]
PRODUCTION GOAL = 1 [t/Year]
RESOURCE PRICE = 1 [$/t]
SPECIFIC OPERATING COST = 0.1 [1/Year]
SPECIFIC PRODUCTION RATE = 1 [t/(t*$*Year)]

Dynamics
profit expectation = RESOURCE PRICE *SPECIFIC PRODUCTION RATE
    *MAX CAPACITY OF RESOURCE [1/Year]
annual production = SPECIFIC PRODUCTION RATE *renewable resource
    *production capital [t/Year]
goal discrepancy = 1 −annual production /PRODUCTION GOAL [1]
net investment = goal discrepancy *annual net profit *INVESTMENT RATE
    /profit expectation [$/Year]
production capital = INTEG (+net investment, INITIAL PRODUCTION CAPITAL) [$]
actual regeneration rate = IF THEN ELSE(renewable resource < EROSION LIMIT, 0,
    NORMAL RESOURCE REGENERATION RATE) [1/Year]
regeneration = actual regeneration rate *renewable resource *(1 -renewable resource
    /MAX CAPACITY OF RESOURCE) [t/Year]
consumption = annual production [t/Year]
renewable resource = INTEG (+regeneration -consumption, INITIAL RESOURCES) [t]
revenue = RESOURCE PRICE *annual production [$/Year]
operating cost = SPECIFIC OPERATING COST *production capital [$/Year]
annual net profit = revenue -operating cost [$/Year]
accumulated net profit = INTEG (annual net profit, 0) [$]

Simulation time parameters
INITIAL TIME = 0 [Year]
FINAL TIME = 100 [Year]
TIME STEP = 0.05 [Year]

Simulation results

Figure Z417b shows the time response of the system for the parameters of the default setting. With the specific rates chosen the system shows profitable exploitation of the resource for the first 60 years. This is followed by a phase of net losses. After about 100 years the renewable resource is exhausted completely.

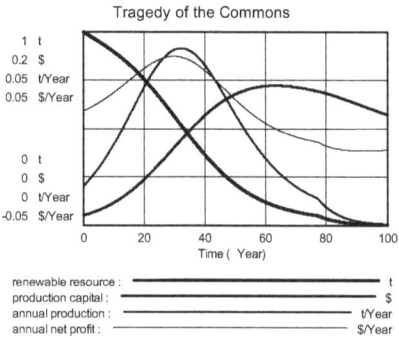

Figure Z417b: Overuse of the renewable resource leads to collapse.

At the beginning of the exploitation and as a result of the (initially high) *annual net profit*, *production capital* increases by *net investment*, thus quickly increasing *annual pro*duction. As long as *annual net profit* is positive, even if the PRODUCTION GOAL is not attained, *production capital* will be increased by *net investment*. *Annual production* and corresponding *proceeds* increase accordingly. But this reduces the amount of *renewable resource* and later also *annual production*. As *production capital* increases, *operating cost* rises accordingly, causing steady decrease of *annual net profit* which eventually goes to zero and then becomes negative. Despite this, production continues until finally the *renewable resource* has disappeared and the resource base is destroyed.

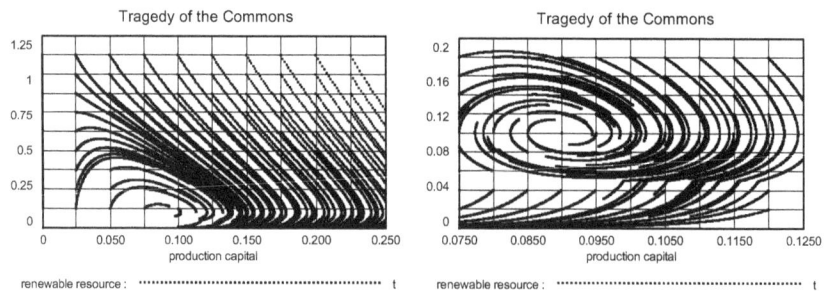

Figure Z417c: Two possibilities of system behavior: oscillatory approach to an equilibrium point with low level of renewable resource, or complete erosion of the resource. The system has three equilibrium points.
Figure Z417d: More exact representation near the stable equilibrium point.

The state trajectories (*renewable resource* vs. *production capital*) in Figure Z417c and d show the behavior of the system for these parameter values even more clearly. The system almost always collapses except for a small (egg-shaped) region around the only non-trivial stable equilibrium point (0.09, 0.1). To be able to use the resource permanently as a commons, strict rules should delimit the use exactly to this region. The system has another two equilibrium points at (0, 0) (no production, no resource) and (0, 1) (no production, undisturbed resource).

If certain system parameters are changed, then the behavior of this system also changes completely and qualitatively. Figures Z417e and f show system trajectories in the state diagram for parameters changed as follows; all other parameters are those of the default setting:
INVESTMENT RATE = 0.25 [1/Year]
PRODUCTION GOAL = 0.01 [t/Year]
SPECIFIC OPERATING COST = 0.25 [1/Year]
SPECIFIC PRODUCTION RATE = 0.5 [t/(t*$*Year)]
The behavior is significantly more complex now. A stable equilibrium point (at 0.0225, 0.887) attracts the state trajectories in the upper range at (approximately for *renewable resource* > 0.75) and in a narrow region at the left hand edge. Trajectories originating from other regions lead to collapse of the system. The system now has a total of five equilibrium points (0, 0), (0, 1), (0.0225, 0.887), (0.1, 0.5) (0.1775, 0.113). The example shows that parameters change can also lead to qualitatively completely different system behavior.

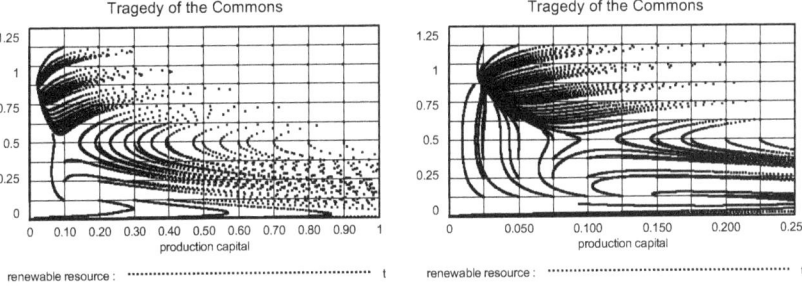

Figure Z417e: State trajectories for changed parameters: The system now has five equilibrium states.
Figure Z417f: Enlarged representation of the left hand region.

If the product (MAX CAPACITY OF RESOURCE * NORMAL RESOURCE REGENERA-TION RATE) is greater than (4 * PRODUCTION GOAL) – i.e. if PRODUCTION GOAL is small compared to the rate of *regeneration* of *renewable resource* – an acceptable permanent solution is possible (the system still has five points of equilibrium). If the PRODUCTION GOAL is too high in comparison to NORMAL RESOURCE REGENERATION RATE, three equilibrium points and corresponding collapse behavior are obtained. The RESOURCE PRICE that can be obtained for the product plays a critical role for system development. If the price is high, this leads to a greater amount of *production capital* and correspondingly greater exploitation.

Exercises

1. Examine the parameter sensitivity with respect in particular to RESOURCE PRICE, SPECIF PRODUCTION RATE, PRODUCTION GOAL, and INVESTMENT RATE.
2. Where are stable equilibrium points located under realistic conditions? Which measures are required if the system is to be used under these conditions?
3. Condense the model mathematically to two differential equations (for *renewable resource* and *production capital*; *accumulated net profit* is calculated only for information purposes and is not relevant to the behavior since it does not have any feedback to the system). Confirm analytically the three or five equilibrium points mentioned above for the parameter combinations stated.
4. Compare the model and its results with model Z418 "Sustainable use of a renewable resource". Determine the reasons for the decisive differences in system behavior.

References

Hardin, G. 1968: The tragedy of the commons. *Science*, vol. 162, pp. 1243-1248.

Z418 Sustainable use of a renewable resource

Simulation task

As model Z417 has shown, the principle of the commons allows sustainable use of renewable resources only if certain constraints are strictly observed – e.g. if population stays constant and each family is allowed to graze only a limited number of cattle on the commons. The authorized number of cattle is often the result of centuries of experience. It represents the knowledge that in the long run exploitation of a renewable resource must match its regeneration. Utilization of renewable resources therefore enforces, first, constraints on selfishness to prevent harm to the community as a whole, and second – in a democracy – just allocation of use rights to all. Our dependence on natural regeneration processes therefore also inevitably has social consequences – even if some are not (yet) willing to recognize this fact.

Sustainable management of renewable resources has been practiced for several centuries as fundamental "principle of sustainability" in the forestry of some countries. In general, all renewable resources should be used according to this principle: not to use more than can regenerate or renew by natural processes. This applies as well to forestry and agriculture as to fishing or hunting, and use of water, air, and soils. Sustainability therefore must not be orientated by reference to (still) available stocks, but must constantly keep in view the regeneration ability of the resource to ensure its permanent regeneration and survival. Exploitation of a resource should remain inherently stable, i.e. sustainability should be automatically assured even without permanent supervision of strict constraints. With simulation models the possibilities can be examined.

Simulation model

Figure Z417a shows the simulation diagram of the model. The model instructions are listed in the following.

The model structure is largely identical to that of model Z417 "Tragedy of the commons": *Renewable resource* grows in a logistic saturation process. The *net investment* in new *production capital* is proportional to *annual net profit* from *revenue* and *operating cost*, but is now controlled (via feedback from *renewable resource*) in such a way that the stock of *renewable resource* that is maintained allows sustainable harvesting with particularly favorable results. The goal for the stock of *renewable resource* corresponds to one half of MAX RESOURCE CAPACITY (cf. maximization of harvest in model Z110 "Logistic growth with stock-dependent harvest" in *System Zoo 1*). If the stock drops below this value, production capital is put out of commission. For a sufficient stock of *renewable resource* and if *annual net profit* is positive, *production capital* is built up further by *net investment.*

Parameters and initial values
INITIAL PRODUCTION CAPITAL = 0.01 [$]
INITIAL RESOURCES = 1 [t]
MAX CAPACITY OF RESOURCE = 1 [t]
RESOURCE PRICE = 1 [$/t]
SPECIFIC PRODUCTION RATE = 1 [t/(t*$*Year)]

INVESTMENT RATE = 0.1 [1/Year]
EROSION LIMIT = 0.05 [t]
NORMAL RESOURCE REGENERATION RATE = 0.1 [1/Year]
SPECIFIC OPERATING COST = 0.1 [1/Year]

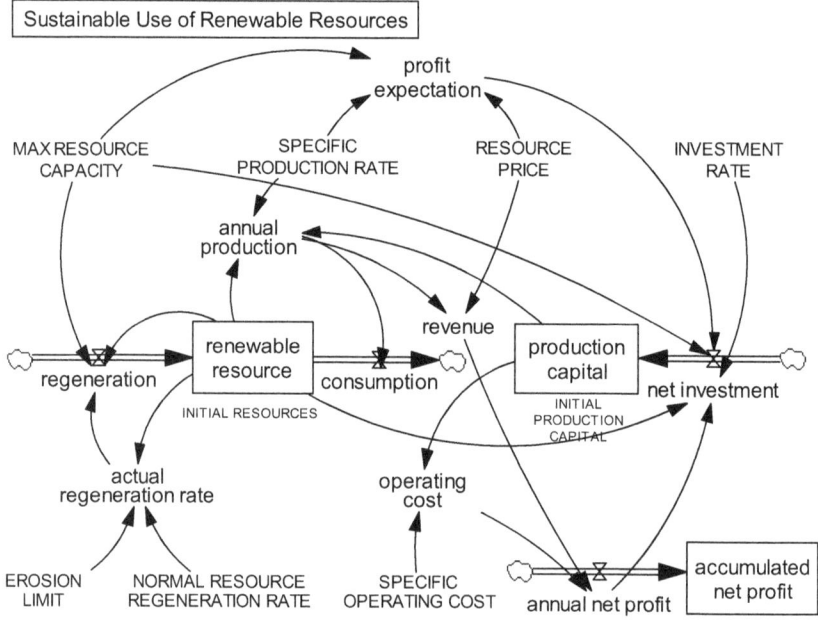

Figure Z418a: Simulation diagram for sustainable use of a renewable resource.

Dynamics
profit expectation = RESOURCE PRICE *SPECIFIC PRODUCTION RATE
        *MAX CAPACITY OF RESOURCE [1/Year]
net investment = (renewable resource /(MAX CAPACITY OF RESOURCE /2) -1)
        *ABS (annual net profit *INVESTMENT RATE /profit expectation) [$/Year]
production capital = INTEG (+net investment, INITIAL PRODUCTION CAPITAL) [$]
annual production = SPECIFIC PRODUCTION RATE *renewable resource
        *production capital [t/Year]
consumption = annual production [t/Year]
actual regeneration rate = IF THEN ELSE (renewable resource < EROSION LIMIT, 0,
        NORMAL RESOURCE REGENERATION RATE) [1/Year]
regeneration = actual regeneration rate *renewable resource *(1 -renewable resource
        /MAX CAPACITY OF RESOURCE) [t/Year]
renewable resource = INTEG (+regeneration -consumption, INITIAL RESOURCES) [t]
revenue = RESOURCE PRICE *annual production [$/Year]
operating cost = SPECIFIC OPERATING COST *production capital [$/Year]
annual net profit = revenue -operating cost [$/Year]
accumulated net profit = INTEG (annual net profit,0) [$]

Simulation time parameters
INITIAL TIME = 0 [Year]
FINAL TIME = 100 [Year]
TIME STEP = 0.05 [Year]

Simulation results

Figure Z418b shows the time behavior of the model for the default parameter settings, using the same scales as Figure Z417b. For the same system parameters as in model Z417 "Tragedy of the commons" development stabilizes after about 70 years at an equilibrium level. Although *annual production* does not reach the peak value of model Z417, it is however sustainable and leads to lasting and constant *annual net profit* and thus altogether to greater *accumulated net profit*.

Without utilization (*consumption*) of the *renewable resource* logistic growth up to MAX RESOURCE CAPACITY would result. With resource utilization *production capital* is built up until *renewable resource* has reached one half of its MAX RESOURCE CAPACITY. Thereafter, the stock of *renewable resource* is maintained at this level. This system structure now leads to constant *annual production* with positive *annual net profit* at constant value of *production capital*.

In the state diagram of Figure Z418c state trajectories move towards a stable equilibrium point (at 0.05, 0.5), if the initial *renewable resource* stock is large enough and the *production capital stock* small enough. Trajectories starting in the outer right hand region eventually reach a state below the EROSION LIMIT, leading to resource collapse. There is a total of four equilibrium points (3 stable, 1 unstable) at (0, 0), (0.05, 0.5), (0.09, 0.1), (0, 1).

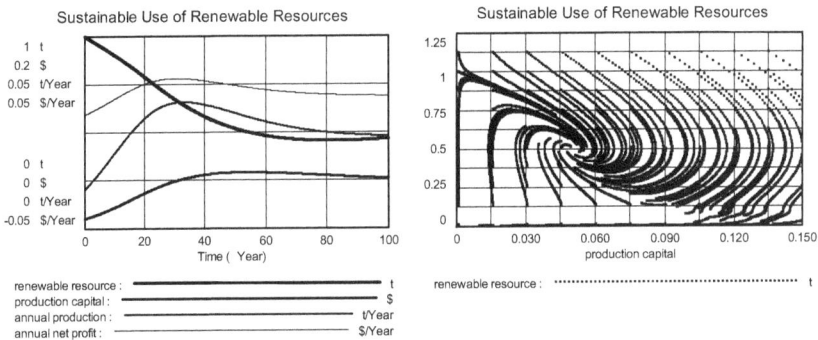

Figure Z418b: Production is sustainable if the resource can permanently regenerate.
Figure Z418c: Collapse is impossible if the state does not approach the erosion limit.

Exercises

1. Compare the behavior of models Z417 and Z418 for identical parameter values. Identify the reasons for the different behavior.
2. Investigate the influence of the different parameters on system development and equilibrium points.

3. Condense the model instructions into two differential equations (leaving out *accumulated net profit*) and determine analytically the location of equilibrium points as function of system parameters. Discuss the result.

4. Find out whether (similar to model Z417) parameter change can cause a "tipping over" to qualitatively different behavior in model Z418.

5. How could this model concept for sustainable management be converted into practice to obtain a reliable and inherently stable system e.g. for common grazing land of a village, or for international fishing regulations?

SYSTEM ZOO PUBLICATIONS

by Hartmut Bossel

All books and CDs are available in internet and local bookstores

English:

System Zoo 1 Simulation Models – Elementary Systems, Physics, Engineering
Books on Demand, Norderstedt, 2007, 184 p. (ISBN 978-3-8334-8422-3)

System Zoo 2 Simulation Models – Climate, Ecosystems, Resources
Books on Demand, Norderstedt, 2007, 204 p. (ISBN 978-3-8334-8423-0)

System Zoo 3 Simulation Models – Economy, Society, Development
Books on Demand, Norderstedt, 2007, 276 p. (ISBN 978-3-8334-8424-7)

Systems and Models – Complexity, Dynamics, Evolution, Sustainability
Books on Demand, Norderstedt, 2007, 372 p. (ISBN 978-3-8334-8121-5)

System Zoo simulation models: *free download of all models from*
Center of Environmental Systems Research, University of Kassel:
http:// www.usf.uni-kassel.de/cesr/ (→ *download* → *software*)
(English versions only; for Vensim PLE)

German:

Systemzoo 1 – Elementarsysteme, Technik und Physik
Books on Demand, Norderstedt 2004, 204 p. (ISBN 3-8334-1239-9)

Systemzoo 2 – Klima, Ökosysteme und Ressourcen
Books on Demand, Norderstedt 2004, 236 p. (ISBN 3-8334-1240-2)

Systemzoo 3 – Wirtschaft, Gesellschaft und Entwicklung
Books on Demand, Norderstedt 2004, 308 p. (ISBN 3-8334-1241-0)

Systeme, Dynamik, Simulation – Modellbildung, Analyse und Simulation komplexer Systeme
Books on Demand, Norderstedt 2004, 400 p. (ISBN 3-8334-0984-3)

Systemzoo CD – 100 Simulationsmodelle
co.Tec Verlag, Rosenheim

Weltmodell World3-03 – Simulationsmodell CD
co.Tec Verlag, Rosenheim (ISBN 3-86563-387-0)